Cassel

UNICODE

UNIVERSAL TELEGRAPHIC PHRASE-BOOK

Cassel

UNICODE
UNIVERSAL TELEGRAPHIC PHRASE-BOOK

ISBN/EAN: 9783741174810

Manufactured in Europe, USA, Canada, Australia, Japa

Cover: Foto ©Andreas Hilbeck / pixelio.de

Manufactured and distributed by brebook publishing software
(www.brebook.com)

Cassel

UNICODE

"UNICODE."

THE

UNIVERSAL TELEGRAPHIC PHRASE-BOOK.

A CODE OF CYPHER WORDS FOR COMMERCIAL,
DOMESTIC, AND FAMILIAR PHRASES IN ORDINARY
USE IN INLAND AND FOREIGN TELEGRAMS.

*WITH A LIST OF PROMINENT COMMERCIAL FIRMS
WHO ARE UNICODE USERS.*

SIXTH EDITION.

CASSELL & COMPANY, LIMITED:

LONDON, PARIS, NEW YORK & MELBOURNE.

1889.

PREFACE.

In first introducing to the public the "Unicode," by means of "The Universal Telegraphic Phrase-Book," * it is well to give a few preliminary explanations and directions.

All the great submarine Telegraph Companies, and almost all foreign countries and colonies, have adopted the word-tariff, or system of charging a certain sum for each word, and Great Britain has practically done so. by the changes effected during the last Parliament. Every person who has heretofore sent telegraphic messages abroad has learned by experience the economy of condensation, and the advantage of the use of a Code known to both sender and receiver. By this means the substance of a message embracing a dozen ordinary words may be conveyed in a single code-word, with a fulness and clearness not to be otherwise attained unless at a prohibitive cost. The same effect is discovered in inland telegraphic communication under the new arrangements. The sixpenny telegram is found, except under special circumstances, to be a misnomer, the unavoidable length of the addresses (where the expense of registering a cypher has not been incurred by the Receiver), and the name of the Sender and Receiver absorbing so many of the twelve words as frequently to leave only two or three available for the text of the telegram. Attention is therefore naturally turned to condensation, and, as a necessary consequence, to coding.

The Code-Book hitherto has been distinguished by two features —a high price and an attempted exclusiveness. The "Unicode" aims at precisely opposite qualities, viz., a low price and a universality of employment, so that not only in all offices, but in clubs, hotels, and private residences copies shall be found and freely used.

* A Pocket Edition of this book, of convenient size, is also published, price 2s. 6d.

An example will best demonstrate the mode of using, and the economy effected. Say the following is the message in full :—

	Smith,			
TO	100, Prince Rupert Road,			
	Shepherd's Bush.			
Jones	dines	with	us	this
evening	and	remains	the	night
				Smith

Here the address and signature take eight words, and the body of the message ten, making eighteen words in all, or six extra to pay for; whereas by using the "Unicode" the message is reduced to ten words, and runs thus :—

	Smith,			
TO	100, Prince Rupert Road,			
	Shepherd's Bush.			
Jones	Coctivus			
		Smith		

But in foreign telegrams the difference is more striking. The rate, for instance, from South America is ten shillings per word, and the following message (without reckoning the address) would cost £3 :—"Order executed before your telegram arrived;" whereas the "Unicode" word "*Obumbro*" would convey the same message at a cost of ten shillings, or a saving of £2 10s.

Many phrases which at first sight would appear too unnecessarily minute, notably in domestic affairs, are purposely inserted as being those which experience shows are in actual daily use, notwithstanding their heavy cost for transmission.

Users of existing codes have constantly experienced difficulty and misunderstanding from the fact that, English words being used for the cyphers, the messages have at times read intelligibly in the ordinary and not the code meaning of the words, and the Receiver has not known which to adopt. This has been entirely obviated in the "Unicode" by exclusively employing for the cyphers Latin words which strictly conform to the regulations of the International Telegraph Conferences held at Paris, London, and Berlin. An equally important point has also been carefully borne in mind. It is generally known that the telegraphic alphabet is composed of three elements: the dot, the dash, and the space. These symbols may with great facility be transposed in transmission, causing words, however dissimilar in ordinary language (such as *fancy* and *pantry*) to be confounded one with another in the process of telegraphy. This compilation, however, has been made under the personal supervision of telegraphic experts of long experience, and it is claimed for it that the cypher words are from their telegraphic construction the least liable to erroneous transmission by the operators.

The cypher words have been arranged alphabetically, and the phrases are likewise so arranged, having regard to what is in each the principal or key-word.

Not the least valuable feature (and it is a novel one) in the present volume is the addition of a list of important firms and establishments in Great Britain, with their registered telegraphic addresses, who will receive telegrams in the "Unicode." This list will be hereafter increased, and for this purpose intimations are invited from those firms at home and abroad who desire their names to be added. These should be sent to the care of the Publishers, and addressed to the Editors of the "Unicode," who will be grateful for any suggestions for improvements and additions.

To allow for the composition of a small private code available only to the individual compilers, and not to be adopted hereafter in the "Unicode" for specific phrases for public use, a few pages with cypher words only have been added, to which phrases may be attached as desired.

NOTE TO SECOND EDITION.

The compilers refer to the following communication which they have received :—

"I have examined from the point of view of a telegraph operator of long standing, several Codes which have been recently published, and I emphatically pronounce the UNICODE to be the only one I have seen where the hand of the expert can be discovered. In one of the other compilations it is claimed as a merit that none of the cypher words exceed five letters. This, however, is a serious blot, and condemns the book in my judgment. In two Codes I observe a free use of manufactured words, and yet, for extra-European correspondence, such words can be and are generally rejected by the Cable Companies. Naturally it is preferable that the operator should transmit messages where the meaning is clear and the words as usual, but as it seems evident that the use of Codes will grow day by day it is to be hoped that the public will continue to employ one so workman-like and systematic as the UNICODE."

October, 1886.

POSSIBLE TRANSFORMATIONS OF TELEGRAPH SIGNALS.

LETTER.	MORSE SIGNAL.	POSSIBLE SUBSTITUTION.			
A	.—	ET			
B	—...	TS	NI	DE	
C	—.—.	NN	TR	TEN	KE
D	—..	TI	NE		
E	.				
F	..—.	IN	ER	UE	
G	——.	ME	TN		
H	....	SE	ES	II	
I	..	EE			
J	.———	EO	ATT	AM	WT
K	—.—	TA	NT	TT	
L	.—..	AI	RE	ED	
M	——	TT			
N	—.	TE			
O	———	TM	MT		
P	.——.	WE	AN	EG	EME
Q	——.—	MA	GT	TK	
R	.—.	AE	EN		
S	...	IE	EI		
T	—				
U	..—	EA	IT		
V	...—	ST	EU	IA	
W	.——	EM	AT		
X	—..—	NA	TU	DT	
Y	—.——	NM	TW	KT	
Z	——..	MI	GE	TD	

REGULÁTIONS AS TO TRANSMISSION OF TELEGRAMS.

—◦◦—

THE rules and regulations which govern the telegraphic communication between various portions of the world are laid down by the International Telegraph Conferences which meet periodically in one or other of the capitals of Europe, and the following gives the effect of those which should be most widely known :—

All telegrams should be legibly written.

Telegrams may be composed of plain language, of code language, or of secret language.

Telegrams in plain language must present a clear meaning in any one of the languages admitted for telegraphic correspondence.

Telegrams in code language must consist of words not exceeding ten letters in length, each of them presenting a clear meaning, but not necessarily any consecutive sense, and belonging to any one or to all of the following languages, but to no other, viz. :—English, French, German, Italian, Dutch, Portuguese, Spanish, and Latin.

Code words containing more than ten letters are charged at cypher rate.

Proper names are not admitted in the text of code messages unless used in their natural sense.

The Company can demand the production of the codes and vocabularies, for the purpose of controlling the execution of the preceding regulations.

Private telegrams composed of secret letters such as a b x y z, are not admitted in extra-European correspondence.

Illegitimate combinations of words contrary to the usage of the language, and abbreviated and wrongly spelt words, are inadmissible.

The maximum length of a single word is fixed at 15 letters in European telegrams, and at 10 letters in extra-European telegrams, any additional letters being counted and charged for as extra words at the rate of 15 or 10 letters respectively to the word.

Any instruction the sender may have to give as to the delivery at destination, prepayment of reply, acknowledgment of receipt, to its being a collated telegram, &c., should be written immediately before the address. These indications may be given in the following abbreviated forms, when they will be counted as one word only :—

RP for	Reply paid.	FS for	To follow.
TC ,,	Collation paid.	RPD ,,	Urgent reply paid.
CR ,,	Acknowledgment of receipt.	PR ,,	Post registered.
PP ,,	Postage paid.	EP ,,	Estaffette paid.
XP ,,	Express paid.		

Any sender may request by writing the instruction : " Télégramme à faire suivre " (*i.e.* to follow) or " (FS) " (which is charged for), immediately before the address, that the terminal office shall cause his telegram to follow the receiver within the limits of Europe.

The charge to the first address only is prepaid, the cost of further transmission being collected on delivery.

INDEX.

UNIVERSAL TELEGRAPHIC PHRASE-BOOK.

Abandon the negotiations . . . . Abactus

(Able)—Am *able* to . . . . . Abazea
Am not *able* to . . . . . . Abdite
Are you *able* to . . . . . Abdixi
Have you been *able* to . . . . Abdo
Shall be *able* to . . . . . Abdomen
Shall not be *able* to . . . . . Abductus

Absence has prevented my earlier reply . Abequito
Can do nothing during *absence* of — . . Abeuntis
Can do nothing in your *absence* . . . Abfore

Accept. (Refer to DECLINE.)
Cannot *accept* less than — . . . . Abhorreo
Do not *accept* . . . . . . Abitio

Acceptance paid away, too late to stop . . Abjecte
Acceptance paid into bank, cannot be with-
drawn . . . . . . Abjectio
Acceptance will be renewed . . . Abjicio
Cannot renew *acceptance* . . . . Abjudico
Have withdrawn *acceptance* . . . Abjuro
Refuse to renew *acceptance* . . . Ablaqueo
Will withdraw *acceptance* on receipt of bank
order for — . . . . . . Ablego

Accident has occurred to train on the — . . Abludo

Met with an *accident*	Abnato
Met with an *accident*, cannot keep appointment	Abnepos
Met with an *accident*, come as quickly as you can	Abneptis
Met with an *accident*, must postpone visit .	Abnodo
Met with an *accident*, must remain here, letter by post	Abnormis
Met with an *accident*, not very serious .	Abolesco
Met with an *accident*, only slight . .	Abolevi
Met with an *accident*, very serious . .	Abolitio
Met with an *accident*, which prevents my leaving	Abominor
Met with an *accident*, will return . .	Aborior
Met with an *accident*, boat upset, all safe .	Abrasi
Met with an *accident*, boat upset, remain here till you come	Abreptus
Met with an *accident*, boat upset, send a change here	Abrogo
Met with an *accident*, carriage upset, not hurt	Abrumpo
Met with an *accident*, carriage upset, slightly hurt	Abruptio
Met with an *accident*, collision, not hurt .	Abscedo
Met with an *accident*, collision, seriously hurt	Abscindo
Met with an *accident*, collision, slightly hurt	Abscisse
Met with an *accident*, come . . .	Absens
Met with an *accident*, but need not come .	Absentia
What is the nature of the *accident* . .	Absilio
When did the *accident* occur . . .	Absolvo
Where did the *accident* occur . . .	Absonus

Account. (Refer to PLACE.)

Account is being made out . . .	Absorbeo
Account is forwarded to-day . . .	Abstineo

Account is overdrawn	**Abstraxi**
Not on my *account*	**Abstrudo**
Not on our *account*	**Absurdus**
Not on your *account*	**Abundo**
Placed to your *account* the sum of — .	**Abusque**

Acknowledge. (Refer to Documents, Letter, Remittance, Telegram.)

Acknowledge by telegram receipt of documents	**Abutor**
Acknowledge by telegram receipt of letter .	**Acacia**
Acknowledge by telegram receipt of remittance	**Academia**
Why have you not *acknowledged* receipt of documents	**Acapna**
Why have you not *acknowledged* receipt of letter	**Acapnon**
Why have you not *acknowledged* receipt of remittance	**Acatium**
Why have you not *acknowledged* receipt of telegram	**Accanto**

Act as for yourself .

Act as for yourself	**Accedo**
Act as you think best	**Accelero**
Act on my letter	**Acceptio**
Act on my previous telegram . . .	**Acceptus**
Act on my previous instructions . . .	**Accerso**
— is empowered to *act* on my (or our) behalf	**Accessio**
You have full powers to *act* . . .	**Accingo**

Address. (Refer to Telegram.)

Registered telegraphic *address* is — . .	**Accio**
What is your registered telegraphic *address*	**Accipio**

Agree to your plans

Agree to your plans	**Accitus**
Agree to your proposals	**Acclamo**
Agree to your request	**Acclinis**
Agree to your terms	**Accola**

May I *agree* to —	Accolens
Will you *agree* to —	Accredo

Agreement is arranged as to terms, but waits

signature	Accresco
Agreement must be sent for our signature .	Accretus
Agreement will be sent for your signature to-day	Accudo
Agreement will be sent on — . . .	Accumbo
Do you confirm the *agreement* . . .	Accumulo
Have not entered into any *agreement* . .	Accursus
Have you entered into any *agreement* . .	Acerbe
They confirm the *agreement* . . .	Acerra
They do not confirm the *agreement* . .	Acetaria
We confirm the *agreement*	Achates
We do not confirm the *agreement* . .	Acheron

Alarmed at not having any news . . .	Achnas
Is there any cause for *alarm* . . .	Acidulus

Allowance is asked for of —	Acinaces
Allowance is too great	Acinosus
Allowance is too small	Aclis
Have had to make a large *allowance* . .	Aconitum
Have had to make a small *allowance* . .	Acopum
What *allowance* is asked for . . .	Acquievi
What *allowance* will you make . . .	Acquiro
What *allowance* would you propose . .	Acredula

(Alongside)—Expected to be *alongside* by — .	Acriter

Alteration cannot be made	Acritude
Alteration has been made as requested .	Acroama
Alteration must be made	Acroasis
Alteration must not be made . . .	Actito
No *alteration* has been made . . .	Actuosus

What *alteration* has been made . .	Actus
What *alteration* is asked for . . .	Actutum
Will make no *alteration* whatever . .	Aculeus
Am entirely in your hands	Acumen
Am I to take charge of — . . .	Acutulus
Am leaving town, but will see you on —.	Adactio
Am quite well, and coming on at once .	Adaggero
Am quite well, and writing by post . .	Adalligo
Am very unwell, come as soon as you can	Adamo
Am very unwell, unable to leave to-day .	Adambulo
Amount is not large enough . . .	Adaperio
Amount is too large	Adaquo
Amount offered is —	Adauctus
Are they good for the *amount* . .	Adaugeo
Cannot obtain increase of *amount* . .	Adauxi
Cannot obtain payment of *amount* . .	Adaxint
Cannot obtain reduction of *amount* . .	Adbibo
Have sent the *amount*	Addecet
Have you sent the *amount*	Addenseo
What *amount* is offered	Addictio
What will it *amount* to	Addisco
Will send the *amount*	Additus
Will you send the *amount*	Addivino
Announcement is confirmed	Addubito
Announcement is contradicted . . .	Adduco
Announcement is doubted	Adegi
Announcement is made officially . . .	Ademptio
Announcement is made privately . . .	Adeps
Announcement is made publicly . . .	Adequito
Announcement is premature . . .	Adesurio
Announcement is quite true	Adesus
Announcement is untrue	Adeuntis

Confirm the *announcement*	Adfrango
Contradict the *announcement* . . .	Adgemo
Annoyed very much at delay	Adhalo
Annoyed very much at refusal . . .	Adhibeo
Annoyed very much at silence . . .	Adhinnio
Annoyed very much at statement . .	Adhortor
Anxious about safety of —	Adhuc
Anxious to have your reply immediately .	Adiantum
Anxious to hear from you about — . .	Adimo
Anticipate little difficulty	Adipatum
Anticipate much difficulty	Adipsos
Anticipate some difficulty	Aditur
Do you *anticipate* any difficulty . . .	Aditurus
Apartments requested are secured at — . .	Adjaceo
Apartments required are engaged, and ready for immediate occupation . . .	Adjugo
Apartments required are engaged, and will be ready for occupation on — . .	Adjunctio
Cannot secure *apartments* you wish . .	Adjunxi
Cannot secure *apartments* you wish, but can get —	Adjutor
Secure *apartments* at —	Adjutrix
Secure a bedroom for me	Admetior
Secure a bedroom and sitting-room . .	Admiror
Secure two bedrooms for — . . .	Admisceo
Secure two bedrooms and sitting-room .	Admistus
Secure three bedrooms	Admitto
Secure three bedrooms and sitting-room .	Admodum
Secure a double-bedded room . , .	Admolior
Secure a double-bedded room and sitting-room	Admoneo
Secure sufficient accommodation for us .	Admordeo

Apply at once for — . . . . . Admorsus
Do not *apply* for — . . . . . Admugio

Appointments. (Refer to CALL, COME, DE-
TAIN, EXPECT, FORGET, MEET, POST-
PONE, TRAIN, WEATHER.)

Appointment has been made for — . . . Admutilo
Cannot attend the *appointment* made . . Adnascor
Cannot keep my *appointment* for — . . Adnavigo
What *appointment* has been made . . Adnoto
Hope to be with you in a few days . Adoptio
Hope to be with you next week . . Adoreus
Hope to be with you on — . . . Adorno
Hope to be with you this evening . . Adpluo
Hope to be with you to-day . . . Adrepo
Hope to be with you to-morrow . . Adscisco
Hope to see you this evening . . Adscitus
Hope to see you to-day . . . . Adsum
. *Hope* to see you to-morrow . . . Advectio
Shall be at home this evening . . Advenio
Shall be at office to-day . . . Adventus
Shall be at office to-morrow . . . Adversor
Shall be at your house this evening . Adverstim.
Shall be at your house to-day . . Advexi
Shall be at your office to-day . . Advigilo
Shall be at your office to-morrow . . Adulator
Shall be in town and will see you on — . Adulor
Shall be in town to-day and will call on
you at — . . . . . . Adumbro·
Shall be in town to-morrow and will call
on you at — . . . . . . Aduncus
Shall expect you this evening . . Advoco
Shall expect you to-day . . . Advolvo
Shall expect you to-morrow . . . Adustio
Wish to see you, and will remain here until— Adytum
B

Wish to see you, and will remain here until you come	Affabre
Wish to see you. Shall I come? Telegraph reply	Affeci
Wish to see you on business . . .	Affectio
Wish to see you on business. Can you come here? Telegraph . . .	Affero
Wish to see you on business. Make an appointment	Affinis
Wish to see you on business. Please wait my arrival	Affirmo
Wish to see you on business. Shall be here until —	Affixi
Wish to see you on business. Shall be with you about —	Affixus
Wish to see you particularly. Can I see you if I call	Afflatus
Wish to see you particularly. Please come here if possible . . .	Affluo
Wish to see you particularly. Please wait my arrival	Affodio
Wish to see you particularly. Shall be here until —	Afforem
Wish to see you particularly. Telegraph time and place	Affrango
Wish to see you particularly. Will be with you about —	Affremo
Wish to see you this evening, call here .	Affrico
Wish to see you this evening, will call on you	Affulgeo
Wish to see you this morning, call here .	Agaricon
Wish to see you this morning, will call on you	Agedum
Wish to see you to-day	Agellus
Wish to see you to-day, telegraph where	Agema
Wish to see you, will be with you at —.	Agesis

Apprehend the worst . . . . . Agger
  There is little cause for *apprehension* . . Aggestus
  There is no cause for *apprehension* . . Aggravo

Arranged everything satisfactorily, return at
  once . . . . Agilis
  Will *arrange* everything to your satisfaction Agilitas

Arrangement has been made . . . . Agitator
  Arrangement has fallen through . . . Agnascor
  Arrangement still under discussion . . Agnatio
  Can you come to any *arrangement* . . Agnos
  Make some definite *arrangement* . . . Agnosco
  What *arrangement* do you propose . . Agnovi
  What *arrangement* is come to . . . Agrarius
  What *arrangement* is suggested . . . Agrestis
  Will not make any *arrangement* . . . Agria

Arrival. (Refer to Goods, Home, Hotels, Train.)
  Arrival is expected on the — . . . Agricola
  Cannot account for non-*arrival*. Will make
    immediate enquiry . . . . Agripeta
  Enquire of agents date of *arrival* . . Agrium
  Enquire of agents date and port of *arrival* . Ahenipes
  Enquire of agents date and port of *arrival*,
    and meet me . . . . . Aizoon
  Glad to hear of your safe *arrival* . . Alabaster
  Shall *arrive* about — . . . . Alacer
  Shall *arrive* and require a conveyance at —. Alacritas
  Shall *arrive* and require a porter at — . Alapa
  Shall *arrive* and require breakfast at — . Alatus
  Shall *arrive* and require dinner at — . . Alauda
  Shall *arrive* and require lunch at — . . Albarius
  Shall *arrive* and require supper at — . . Albesco
  Shall *arrive* and require tea — . . . Albor
  Arrived here after a very bad passage . . Albumen

B 2

Arrived here after a very good passage .	Alburnum
Arrived here, all in good order . . .	Alcedo
Arrived here all well	Algensis
Arrived here all well, health much improved.	Algidus
Arrived here all well, very tired . . .	Algor
Arrived here all well, I will write to-day .	Algosus
Arrived here all well, I will write to-morrow	Alias
Arrived here all well, leave again to-day .	Alica
Arrived here all well, leave again to-morrow	Alicubi
Arrived here all well, leave for — . .	Alienatio
Arrived here all well, leave for home — .	Alienus
Arrived here all well, leave for home to-day	Alifer
Arrived here all well, leave for home to-morrow	Alimon
Arrived here all well, meet me at — . .	Alioqui
Arrived here all well, remain to-night . .	Aliorsum
Arrived here all well, remain until —. .	Altitudo
Arrived here all well, wait my arrival . .	Altrix
Arrived here unwell, meet me at — . .	Alveare
Ascertain the reason and telegraph at once .	Alveolus
Ascertain the reason and write at once .	Alveus
Assistance is not required	Alum
Assistance is urgently required . . .	Amabilis
Ask for what *assistance* you require . .	Amando
Cannot render any *assistance* . . .	Amarus
Will give all the *assistance* in our power .	Amasius
(Avoid)—Do your best to *avoid* — . . .	Amator
Do your utmost to *avoid* any unpleasantness	Amatrix
Bank rate has been raised —	Ambedo
—— ¼ per cent.	Ambesus
—— ½ per cent.	Ambique
—— ¾ per cent.	Ambio
—— 1 per cent.	Ambustio

Bank rate has been reduced — . . .	Amellus
—— ¼ per cent.	Amentia
—— ½ per cent.	Amerina
—— ¾ per cent.	Amicitia
—— 1 per cent.	Amiculum

Bankruptcy petition has been filed by —. .	Amissio
Bankruptcy proceedings have been taken against —	Ammium

Bills. (Refer to ACCEPTANCE.)

Bills of Lading are not yet made out . .	Ammonis
Bills of Lading are sent by this mail . .	Amnicola
Bills of Lading have not been endorsed .	Amolior
Bills of Lading will be sent by next mail .	Amomis
Have you sent *Bills of Lading* . . .	Ampelos
How are *Bills of Lading* forwarded . .	Amphora
How are *Bills of Lading* made out . .	Amplio
Send *Bills of Lading* immediately . .	Amplius

(Births)—Confined this morning, *Boy*, both doing well	Ampulla
Confined this morning, *Boy*, dead, Mother well	Amputo
Confined this morning, *Girl*, both doing well	Amuletum
Confined this morning, *Girl*, dead, Mother well	Amusium
Confined to-day, Baby dead, Mother fairly well	Amussis
Confined to-day, Baby dead, Mother weak .	Amygdala
Confined to-day, Baby dead, Mother very weak	Amylon
Confined to-day, Baby dead, Mother fairly well, will telegraph again	Amystis

Confined to-day, Baby dead, Mother weak, will telegraph again	Anatinus
Confined to-day, Baby dead, Mother very weak, will telegraph again . . .	Anceps
Confined to-day, *Twins*, both *boys*, all well .	Ancilium
Confined to-day, *Twins*, both *girls*, all well .	Ancilla
Confined to-day, *Twins, boy and girl*, all well	Ancon
Confined to-day, *Twins*, one alive, a *boy*, Mother well.	Andabata
Confined to-day, *Twins*, one alive, a *boy*, Mother weak	Andron
Confined to-day, *Twins*, one alive, a *boy*, Mother not expected to live . . .	Anellus
Confined to-day, *Twins*, one alive, a *girl*, Mother well	Anemone
Confined to-day, *Twins*, one alive, a *girl*, Mother weak	Anethum
Confined to-day, *Twins*, one alive, a *girl*, Mother not expected to live . . .	Angina
Confined to-day, *Twins*, both dead, Mother well	Anguifer
Confined to-day, *Twins*, both dead, Mother weak	Anguinus
Confined to-day, *Twins*, both dead, Mother not expected to live	Anguipes
Confined yesterday, *Boy*, both doing well .	Anguis
Confined yesterday, *Boy*, dead, Mother well.	Angulus
Confined yesterday, *Boy*, dead, Mother fairly well	Angustia
Confined yesterday, *Girl*, both doing well .	Anhelo
Confined yesterday, *Girl*, dead, Mother well	Anicetum
Confined yesterday, *Girl*, dead, Mother fairly well	Aniciana
Confined yesterday, *Twins*, both *boys*, all well	Anilis
Confined yesterday, *Twins*, both *girls*, all well	Animalis

Confined yesterday, *Twins, boy and girl*, all
    well . . . . Animatio

Confined yesterday, *Twins*, one alive, a *boy*,
    Mother well. . . . . . Animor

Confined yesterday, *Twins*, one alive, a *boy*,
    Mother weak . . . . . Animosus

Confined yesterday, *Twins*, one alive, a *boy*,
    Mother not expected to live . . . Anisum

Confined yesterday, *Twins*, one alive, a *girl*,
    Mother well . . . . . Annalis

Confined yesterday, *Twins*, one alive, a *girl*,
    Mother weak . . . . . Annavigo

Confined yesterday, *Twins*, one alive, a *girl*,
    Mother not expected to live . . . Annexus

Confined yesterday, *Twins*, both dead, Mother
    well . . . . Annifer

Confined yesterday, *Twins*, both dead, Mother
    weak . . . . Annona

Confined yesterday, *Twins*, both dead, Mother
    not expected to live . . . . Annosus

Book is not yet published . . . . Annumero
Book is not published by us . . . Annuntio
Book is out of print . . . . . Anodyna
The published price of the *book* is — . . Anonium
Book will be published about —. . . Anormis
Last edition of *book* completely sold out . Ansatus
New edition of *book* will be ready — . . Anteago

Bring home or order from— . . . . Antecedo
Bring home or order from the Stores — . Antefero
Bring home or send at once — . . . Antenna
Bring home with you — . . . . Antepono
Bring home with you from — . . . Antequam
Bring home with you from the Stores — . Antesto

Bring some fish with you to-day . . .	Antetuli
Bring some fruit with you to-day . .	Antevolo
Bring some game with you to-day . .	Anthedon
Bring with you when next you come . .	Anthera

Business. (Refer to HEALTH.)

Business is suspended on account of holidays	Anthrax
Business will be entertained . . .	Anticipo
Cannot be at *business* to-day . . .	Anticus
Cannot be at *business* to-day; am suffering from an attack of —	Antidotum
Cannot be at *business* to-day; bring letters, &c., here	Antistes
Cannot be at *business* to-day; send anything requiring my attention here . . .	Antlia
Cannot be at *business* to-day; send clerk with letters, telegrams, &c. . . .	Antrum
Cannot be at *business* to-day; send messenger with letters, &c.	Anxietas
Cannot be at *business* to-day; too unwell .	Anxifer
Cannot be at *business* to-day; unnecessary to send messenger, but post letters . .	Anxiferum
Cannot be at *business* to-day until late .	Apage
Cannot be at *business* for a few days . .	Apagesis
Cannot be at *business* for a few days; letter by post	Apathes
Do not do the *business*	Apecula
Do you consider the *business* sound . .	Aperio
When will *business* be concluded . .	Apertura

Buy for me on best terms . . . . Apexabo

Can *buy* at —	Aphaca
Can *buy* more on same terms . . .	Aplustre
Can you *buy* —	Apocynon
Can you *buy* at	Apolecti

Can you *buy* more on same terms . .	Apologus
Cannot *buy* at —	Apostema
Cannot *buy* more	Apotheca
Do not *buy*	Appareo
Do not *buy* any more.	Appendix
What price can you *buy* at	Appendo
What price did you *buy* at	Appensus
What quantity can you *buy* . . .	Appeto
What quantity did you *buy* . . .	Appiana

Call at once upon —	Applaudo
Call at post office for letters . . .	Apporto
Call at this address	Appositus
Call here on your way to business . .	Appotus
Call here on your way from business . .	Apprime
Call here to-day if possible . . .	Apprimus
Call here to-day, without fail . . .	Appulsus
Call on me at my office	Aprilis
Call on me at my office at once . . .	Apronia
Do not *call* upon	Aprugnus
Calling on you to-day	Apsis
Calling on you to-day on an important matter	Apsyctos
Calling on you to-day on important business	Apyrinus
Calling on you to-day with reference to — .	Apyrum
Calling on you to-day as desired . .	Aqua

Cancel. (Refer to TELEGRAM.)

Cancel my previous telegram, and substitute following	Aquator
Cancel orders already given respecting — .	Aquosus
Cancel orders at any cost, reply by telegram	Arabilis
Cancel orders if not already attended to .	Aranea
Cancel orders if not already attended to, letter follows	Araneola
Cancel orders in my telegram . . .	Aratrum

Cancel orders in my telegram, letter by post	Arbiter
Cancel orders in my letter	Arbitror
Cancel orders previously sent, and substitute	
following —	Arboreus
Cancel orders, wait further instructions .	Arbusto
Cannot *cancel* orders, already attended to .	Arbustum
Cannot *cancel* orders already attended to,	
letter follows	Arbuteus

Carriage must be charged forward . . .	Arcanus
Carriage must be prepaid	Archon
Carriage must be sent for me . . .	Arctos
Carriage need not be sent for me . .	Arcturus

Charge has been made in error . . .	Arcuatim
Charge has been withdrawn . . .	Ardelio
Am I to take *charge* of —	Ardenter
Take charge of —	Ardesco
Take charge of everything	Arduitas
Take charge of everything until my arrival .	Arefacio
Whom do you wish to take *charge* of — .	Arenosus
Charges must be paid by — . . .	Arenula
Shall I pay *charges*	Areola
Who will pay the *charges*	Argema

Cheque. (Refer to Money, Remittance.)	
Cheque is sent to-day	Argemone
Cheque has been duly paid	Argentum
Cheque has been lost. Stop payment . .	Argilla
Cheque has been presented and paid . .	Argus
Cheque has been presented and returned	
marked —	Arguto
Cheque returned unpaid, send cash by return	
of post	Argyritis
Cheque will be sent to-morrow . . .	Ariditas
Has *cheque* been paid	Arieto

Has *cheque* been sent . . . . . Aritudo
How was *cheque* sent . . . . Armarium
Make *cheque* payable to bearer . . . Armatura
Make *cheque* payable to our order . . Armifer
Send uncrossed *cheque* payable to bearer . Aroma

Christmas greetings to you . . . . Arrectus

Circumstances beyond my control compel me to
    decline . . . . . . Arrexi
Circumstances beyond my control prevent my
    accepting . . . . . . Arrhabo
Under no *circumstances* . . . . Arripio
What are the *circumstances* of the case . Arrisor

Claim has been allowed . . . . . Arsurus
Claim has been disallowed . . . . Artemon
Send us particulars of the *claim* . . . Arteria

Come as soon as you can . . . . Articulo
Can you *come* here . . . . . Arvalis
Cannot *come* to-day . . . . . Arvina
Cannot *come* to-night,[accept my apologies . Aruspex
Cannot *come* until —. . . . . Asarum
Do not *come* . . . . . . Asbolus
Do not *come*, am leaving for home .· . Ascendo
Do not *come*, reasons by letter . . . Ascopera
Do not *come* until you get my letter . . Ascribo
Do not *come* until — . . . . Asellus
Glad to hear you are *coming* . . . Asinarius
Glad to hear you are *coming*, will meet
    you . . . . Asotus

Commence as soon as possible . . `. Aspecto
Cannot *commence* before . . . . Aspergo
When do you *commence* . . . . Aspernor

Commission must be provided for of — . . Aspexi
Commission will be allowed of — . . Aspicio
What *commission* will be allowed . . Aspiro

Communication by telephone is interrupted . Asplenon
Address all *communications* to — . . Asporto
In *communication* with — . . . . Aspretum
In *communication* respecting the — . . Assecla

Compliments of the season . . . . Assector
Compliments of the season to all from all here Assensio

Comply with their requirements as far as you can Assequor
Comply with their requirements in all respects Assero
Comply with their requirements under protest Assessor
Cannot *comply* with your wishes . . Assevero
Will *comply* with your wishes . . . Assicco

Compromise, if you think it desirable . . Assideo
Compromise on the terms indicated . . Assigno
Compromise upon any terms . . . Assimulo
Do not *compromise* . . . . . Assisto

Concerts. (See THEATRES AND CONCERTS.)

Conclude negotiations at once, or break off . Associo
Have you come to any *conclusion* . . Assolet

Condole with you in our mutual great loss . Assuesco
Condole with you in your great loss . . Assulose

Confidential agent will be sent on our behalf . Assulto
Following is strictly *confidential* . . Assyrius

Congratulate you on the birth of a daughter . Astaphis
Congratulate you on the birth of a son . Asterion
Congratulate you on the happy event . . Asterno

Congratulate you on the well-merited honour	Astituo
Congratulate you on your appointment	Astrepo
Congratulate you on your birthday	Astrifer
Congratulate you on your good fortune	Astringo
Congratulate you on your marriage	Astrum
Congratulate you on your promotion	Astruxi
Congratulate you on your safe arrival	Asturco
Congratulate you on your success	Atavus

Conjoint action is advisable	Athara
Are you acting in *conjunction*	Atheroma

Consequences will be very serious	Athleta
It is of great *consequence*	Atocium
It is of no *consequence*	Atricolor

Consideration must be given to the proposal	Atriolum
Cannot reply without further *consideration*	Atriplex

Consignment duly despatched	Atrium
Consignment duly received	Atrophia
How is it *consigned*	Attactus

Consult some authority in the matter	Attagen
Consult some authority in the matter and let me know result	Attestor
Consult your business agent	Attexo
Consult your business agent and let me know result	Atthis
Consult your friends	Attollo
Consult your friends and let me know result	Attonite
Consult your legal adviser	Attonsus
Consult your legal adviser and let me know result	Attractus
Consult your stockbroker	Attraho
Consult your stockbroker and let me know result	Attremo

Have consulted business agent, who says — . . Attritus
Have consulted business agent, will post par-
    ticulars . . . . . . Auctio
Have consulted friends, who say — . . Aucupium
Have consulted friends, will post you parti-
    culars . . . . Audacia
Have consulted legal adviser, who says — . Audax
Have consulted legal adviser, will post par-
    ticulars . . . . . . Audenter
Have consulted stockbroker, who says — . Auditio
Have consulted stockbroker, will post par-
    ticulars . . . . . . Aufero
Cannot attend the *consultation* . . . Aufugio
When is the *consultation* fixed for . . Augesco

Continue to advise fully by letter . . . Augmen
Continue to advise fully by telegram . . Augurium
Continue the negotiation . . . . Auletes

(Convenient)—It is quite *convenient* . . . Aureolus
It is not *convenient* . . . . . Auresco
Whichever is most *convenient* . . . Auricula
Will it be *convenient* to . . . . Aurifex

Cost is estimated at — . . . . . Auriga
Cost must not exceed — . . . . Auritus
Cannot estimate *cost* . . . . . Aurora

Countermand the order at once . . . Ausculto
It has been *countermanded* . . . Auspicor

Country post has been delayed . . . Austerus
Will reply on my return from the *country* . Avaritia
Will telegraph on arrival in the *country* . Avarus
Will write on arrival in the *country* . . Aveho
— is in the *country*, will communicate with
    him and then reply . . . . . Caballus

Damage is serious . . . . . . Cacabus
Damage is slight . . . . . Cachexia
How long will it take to repair *damage* . Cachinno
How was *damage* caused . . . . Cachla
What is amount of *damage* done . . . Cadaver

Date of last letter is — . . . . . Cadivus
What *date* do you arrive . . . . Cadmites
What *date* do you leave . . . . Caduceum
What *date* does trial commence . . . Caelamen
What *date* does your leave commence . . Caelum
What *date* does your leave terminate . . Calcitro
What *date* was your last letter . . . Calefacio
What *date* was your last telegram . . Caligo

Day after to-morrow . . . . . Callidus
How many *days* can you allow . . . Calthula
In the course of the *day* . . . . Calvaria
In the course of the last few *days* . . Calvatus
In the course of the next few *days* . . Calyx

Dealt with them for many years . . . Camella

(Deaths)—When did he (or she) *die* . . Camera
Died suddenly, come at once . . . Caminor
Died suddenly, do not come, will write you. Cancello
Died suddenly, require instructions . . Cancer
Died suddenly, will write particulars . . Candela
Died to-day, come at once . . . . Candesco
Died to-day, do not come, will write . . Canicula
Died to-day, will write particulars . . Canopus
Died yesterday, come at once . . . Cantamen
Died yesterday, do not come, will write . Cantator
Died yesterday, will write particulars . . Canticum
Baby died to-day, particulars by letter . Capedo
Baby died yesterday, particulars by letter . Capella

Daughter died to-day, particulars by letter .	Capistro
Daughter died yesterday, particulars by letter	Capitium
Father died to-day, come at once . .	Capnitis
Father died to-day, do not come, will write	
you	Cappa
Father died yesterday, particulars by letter .	Capsula
Father died to-day, particulars by letter .	Captatio
Grandchild died to-day, will write you .	Captiose
Grandchild died yesterday, will write you .	Captus
Grandfather died to-day, will write you .	Carbasus
Grandfather died yesterday, will write you .	Carbo
Grandmother died to-day, will write you .	Cardisce
Grandmother died yesterday, will write you .	Carectum
Husband died to-day, come at once . .	Carmino
Husband died to-day, do not come, will	
write you	Carnifex
Husband died to-day, particulars by letter .	Carphos
Husband died yesterday, particulars by letter	Carpinus
Mother died to-day, come at once . .	Carptim
Mother died to-day, do not come, will write	
you	Carruca
Mother died to-day, particulars by letter .	Caryotis
Mother died yesterday, particulars by letter.	Casia
Sister died to-day, come at once	Cassida
Sister died to-day, do not come, will write you	Castanea
Sister died to-day, particulars by letter .	Castigo
Sister died yesterday, particulars by letter .	Castus
Son died to-day, come at once . . .	Catasta
Son died to-day, do not come, will write you	Cathedra
Son died to-day, particulars by letter . .	Catillus
Son died yesterday, particulars by letter .	Caucon
Wife died to-day, come at once . . .	Caudex
Wife died to-day, do not come, will write you	Caupona
Wife died to-day, particulars by letter .	Causatio
Wife died yesterday, particulars by letter .	Cautim
Take charge of all *effects*, letter by post .	Cavator

Funeral takes place at —	Caveo
Funeral takes place on the — . . .	Cedo
Funeral takes place and trust you will come on the —	Cedratus
When does *funeral* take place . . .	Cedrium
Send me a lock of his (or her) *hair* . .	Celeber
Will cannot be found	Cellula
Will cannot be found, can you give any information	Celsus
Has made a *Will* which is now in the custody of —	Cenchris
Has made no *Will*	Censio

Decision. (Refer to FINAL.)

Decision may be expected about — . .	Centrum
When will it be *decided*	Centuria

Decline to accept . . . . . . Cepa

Decline to accept, except on terms proposed.	Cepetum
Decline to accept, except on terms already mentioned	Cephalus
Decline to accept on any conditions . .	Cepphus
Decline to accept on terms mentioned . .	Cepurica
Decline to accept the responsibility . .	Ceraria
Decline to accept under any circumstances .	Cerastes

Deduction proposed is agreed to . . .	Ceratium
No *deduction* can be allowed . . .	Cerberus
Will pay on *deduction*	Cerealis

Delay departure	Cerifico
Delay departure until you receive my letter	Cerno
Delay departure until you hear again . .	Ceroma
Cannot be longer *delayed*	Cerritus
Further *delay* is unnecessary . . .	Certamen
Is any *delay* likely to occur . . .	Cervix

o

It was unavoidably *delayed*   .   .   .   Cestus
There will be considerable *delay* .   .   .   Cetarius
There will be some *delay* .   .   .   .   Chalcis

Deliver only against payment   .   .   .   Chaos
Can only *deliver* on prepayment .   .   .   Charta
Cannot be *delivered* owing to insufficient
address .   .   .   .   .   .   Chelonia
Cannot *deliver* the goods by the time
named .   .   .   .   .   .   Chelys
Cannot *deliver* the goods until —   .   .   Chersos
Do not *deliver* until you receive instruc-
tions .   .   .   .   .   .   Chia
Delivery can be made at once   .   .   .   Chimaera
Delivery can be made in —   .   .   .   Chlamys

Departure.   (Refer to DELAY, DETAINED,
EMBARKING, LEAVING, PASSAGE, TRAIN,
WEATHER.)
Departure postponed .   .   .   .   .   Chlorion
Departure postponed until —   .   .   .   Chorus
Departure postponed until next mail .   .   Chreston
Departure postponed for a few days   .   .   Chroma
Departure postponed for one month   .   .   Chrysos
Departure postponed for one month, letter
follows   .   .   .   .   .   .   Cibalis
Departure postponed for six weeks   .   .   Ciborium
Departure postponed for six weeks, letter
follows   .   .   .   .   .   .   Cicatrix
Departure postponed for two months .   .   Cicera
Departure postponed for two months, letter
follows   .   .   .   .   .   .   Ciconia

Describe exactly what you want   .   .   .   Cimex
Describe the position fully .   .   .   .   Cimolia

Send us detailed *description* . . . Cinctura
We must have better *description* . . Cingulum

Despatch is of utmost importance . . . Ciniflo
Cannot be *despatched* before — . . . Circa
When will you *despatch* . . . . Circiter
Will be *despatched* by — . . . . Circulus
Has been already *despatched* . . . Circumdo

Detained here . . . . . . Cisium
Detained here by contrary winds, will advise
    departure . . . . . . Cista
Detained here by heavy gale, will advise
    departure . . . . . . Cistifer
Detained here, cannot return to-day . . Citerior
Detained here, cannot return to-night . . Cithara
Detained here, do not expect me . . . Citimus
Detained here, do not expect me until — . Citrum
Detained here, do not wait . . . Civicus
Detained here, do not wait, will follow . Clamito
Detained here, expect me about — . . Clamor
Detained here, expect to leave about — . Clango
Detained here owing to private affairs . Clathro
Detained here, shall be with you later . . Clavulus
Detained here, shall dine at the club . . Clemens
Detained here, shall not be home to dinner . Clepere
Detained here, will return — . . . Clibanus
Detained here, will return to-night . . Clinamen
Detained here, will return to-morrow . . Clinice
Detained here for a few days . . . Clivina
Detained here for a few days, letter by post. Clivosus
Detained here until next mail . . . Cloaca
Detained here until next mail, letter by post Cludo

Difficulties. (Refer to ANTICIPATE.)
Difficult to carry out your requirements . Clunis
  c 2

Difficulties exist, but they may be overcome	Clypeus
Difficulties exist which cannot be overcome .	Coactio
Do any *difficulties* exist	Coactus
Diligence is of greatest importance . . .	Coaggero
Diligence shall be exercised . . .	Coagulum
Dimensions in detail must be sent us . .	Coalesco

(Dinner Engagements)—*Accept* your invitation

to *dine*	Coarcto
Accept your invitation to *dine*, and will call at —	Coaxatio
Accept your invitation to *dine*, and will wait here for you	Coccinus
— *dines* with us this evening . . .	Cocles
— *dines* with us this evening, and remains the night	Coctivus
— *dines* with us this evening, but leaves early	Codex
— *dines* with us this evening, shall leave it entirely to you	Coetus
— *dines* with us this evening, will be glad if you will join us	Cognatus
Dining out this evening, do not expect me until —	Cognomen
Dining out this evening, send my dress clothes here	Cognosco
Dining out this evening, send my dress clothes to —	Cohors
Dining out this evening, will join you at —	Cohortor
Dining out this evening with — . .	Colaphus
Received *invitation* to dine with — . .	Collabor
Received *invitation* to dine and Theatre this evening with —	Collaria
Received *invitation* to dine and Theatre, can you come	Collaudo

Received *invitation* to dine and Theatre,
shall I accept . . . . . Collectio
Will dine with you on the — . . . Collido .
Will dine with you to-day. . . . Colloco
Will dine with you to-day, and wait your
arrival at — . . . . . Collum
Will dine with you to-day, and call at — . Collybus
Will dine with you to-morrow . . . Collyra
Will dine with you to-morrow, and wait for you Colonia
Will dine with you to-morrow, and call for
you . . . . Colorate
Will dine with you on Monday . . . Colossus
Will dine with you on Tuesday . . . Columba
Will dine with you on Wednesday . . Colurnus
Will dine with you on Thursday . . Coluthea
Will dine with you on Friday . . . Comatus
Will dine with you on Saturday. . . Comicus
Will dine with you on Sunday . . . Comitium
Will you dine with me — . . . Commadeo
Will you dine with me to-day at — . . Commigro
Will you dine with me to-day, will call on
you at — . . . . . . Committo
Will you dine with me to-day at the Club
at — . . . . Commode
Will you dine with me to-day, and call for
me at — . . . . . . Commorit
Will you dine with me to-day here at — . Commuto
Will you dine with me to-day with a few
friends, at — . . . . . Compactio
Will you dine with me to-morrow at — . Compago
Will you dine with me to-morrow at the
Club, at — . . . . . . Compasco
Will you dine with me to-morrow here at — Compedis
Will you dine with me to-morrow with a
few friends, at —. . . . . Compello

Will you dine with me to-morrow, will call on you at —	Compingo
Will you dine with me to-morrow, and call for me at —.	Compitum
Will you dine with me on Monday at — .	Complano
Will you dine with me on Tuesday at — .	Complico
Will you dine with me on Wednesday at —	Compos
Will you dine with me on Thursday at — .	Comprimo
Will you dine with me on Friday at — .	Comptus
Will you dine with me on Saturday at — .	Concivi
Will you dine with me on Sunday at —. .	Conclamo
Will you dine with me on the — . .	Concluse
Directions have been given	Concolor
What *directions* have been given . .	Concoquo
Disbursements amount to —	Conculco
Send full account of *disbursements* . .	Concumbo
What do *disbursements* amount to . .	Condemno
Dispatch. (See DESPATCH.)	
Dispose of it as you please	Condoleo
Please hold at *disposal* of — . . .	Condris
Would they be *disposed* to — . . .	Conduxi
Dispute has arisen between —. . . .	Condylus
There is no *dispute*	Confero
What is the cause of the *dispute*. . .	Confisus
(Distance)—What is the *distance* . . .	Conformo
Do as you propose	Confusus
Do as you think best	Congelo
Do your utmost for our joint benefit . .	Congeries
Do your utmost in the matter . . .	Congius

Do your utmost on my behalf . . .	Conglobo
Have *done* as you requested . . .	Congruo

Doctor. (Refer to HEALTH, MEDICINE.)

Documents. (Refer to ACKNOWLEDGE.)

Documents are signed. What is to be done with them	Congruum
Documents for your signature have been forwarded	Conifer
Documents for your signature have been forwarded by registered letter . . .	Coniferum
Documents signed and returned you . .	Conisso
Documents signed and posted to-day . .	Conjugo
Documents will be signed — . . .	Conjux
Documents will be signed and sent you — .	Connexus

(Effort)—Every *effort* has been made . .	Conopeum

(Elsewhere)—Shall I try *elsewhere* . .	Conquiro

Embarking on board the —	Conscius
When do you *embark*	Conseco

Empower you to act for me	Consepio
— is *empowered* to act	Consisto
Is he *empowered* to act	Consolor
— is not *empowered* to act	Consors.

(Enclosure)—Send for *enclosure* to-day to — .	Conspuo
Send for *enclosure* to-morrow to — . .	Consulto

Enquire fully and report by letter . . .	Contagio
Enquire fully and report by telegram . .	Contendo
Enquire of agents respecting — . . .	Continuo

Enquire of carriers	Contorsi
Enquire of railway company . . .	Contra
Have made enquiries	Contremo
Have made enquiries and will post you result	Contumax
Have made enquiries of agents, who say —	Contundo
Have made enquiries of carriers, who say —	Convecto
Have made enquiries of railway company, who say —	Converse
Erection is proceeding rapidly . . .	Convexi
Erection is proceeding slowly . . .	Conviva
How is *erection* proceeding . . .	Convoco
How long will *erection* take . . .	Convolvo
Error has arisen	Conus
It is a clerical *error*	Coopto
(Essential)—Do you consider it *essential* . .	Coquino
It is most *essential*	Coram
It is not *essential*	Corbula
Estimates cannot be furnished before — . .	Corculus
Estimate has been accepted . . .	Cordyla
Estimate has been rejected	Cornipes
Estimated loss is —	Corolla
Estimated profit is —	Corpus
Estimates are high	Corrado
Estimates are low	Corsa
Estimates will be sent on by — . . .	Cosmeta
Can you give us an *estimate* . . .	Cotoneum
What is the *estimated* value of — . .	Cottana
Everything arranged	Coturnix
Everything left entirely to you . . .	Crabro

Examination will take place on the — . .	Crapula
Failed in the *Examination*	Crassus
Passed successfully in the *Examination* .	Crates
Exception cannot be made	Creatrix
Exception will be made	Credulus
Executors appointed are —	Crematio
Who are appointed *executors* . . .	Cremo

Expect. (Refer to HEALTH.)

Expect to be with you —	Crepida
Expect to be with you this evening . .	Crepitus
Expect to be with you to-morrow . .	Cribrum
Expect to be with you in a few days . .	Crispus
Expect to be with you in time for — . .	Croceus
Am daily *expecting* to receive information .	Crocota
Have been *expecting* letter from my husband. Telegraph how he is	Crudesco
Have been *expecting* letter from my wife. Telegraph how she is	Crudus
Have been *expecting* letter. Telegraph health of —	Crumena
Have been *expecting* letter. Telegraph how you are getting on.	Crusta
Have been *expecting* letter. Telegraph news	Crux
Have been *expecting* to hear from you. .	Crypta
Have been *expecting* to hear from you. Are you well	Cubatus
Have been *expecting* to hear from you. Very anxious	Cucuma

Expenses to be paid by —	Culearis
At whose *expense*	Culminia
Do your utmost to avoid unnecessary *expense*	Cultor

No *expense* must be incurred . . . Cultrix
Will bear the *expense*. . . . . Cumulate

Experiments have been successful . . . Cuneo
Experiments have not been successful . . Cunque
Report result of *experiments* . . . Cupa

Explanation impossible by telegraph . . Curculio
Explanation sent by letter . . . . Curiosus

Facilitate matters as much as you can . . Cursito
Do nothing to *facilitate* them . . . Curvatio
Shall be happy to offer any *facilities* . . Custodia
What *facilities* have they for — . . Cuticula
What *facilities* have you for — . . Cyamos

Fault has been remedied . . . . Cybium
Fault is entirely theirs . . . . Cyminum
Fault is ours . . . . . . Cynicus
Fault is not ours . . . . . Cynomyla
Fault is yours . . . . . . Cynosura
What is the nature of the *fault* . . . Cyperis
Was *faulty* when it reached us . . . Dabula

Final decision come to is — . . . Dactylis
Consider the decision *final* . . . . Daemon
Is your decision to be considered *final* . . Damnatio

Find out all you can, and report . . . Damnose
Where can we *find* — . . . Danista
Have *found* what is wanted . . . Dapalis

Finish as quickly as possible . . . . Daphne
Finish work at any cost . . . . Dartos
Expect to be *finished* on or about — . . Dasypus
How soon can it be *finished* . . . Datarius

Fire broke out — . . . . . . Dealbo
   Fire broke out this morning . . . Decanto
   Fire broke out to-day, little damage, working
      as usual . . . . . . Decennis
   Fire broke out to-day, great damage . . Decessor
   Fire broke out to-day, great damage, work
      stopped . . . . . . Decidium
   Fire broke out to-day, come at once . . Decoctor
   Fire broke out last night . . . . Decollo
   Fire broke out last night, little damage . Decresco
   Fire broke out last night, great damage . Decuria
   Fire broke out last night, still burning . Decursus
   Fire broke out here, place completely destroyed Defamo
   Fire broke out close here, premises in danger Defectio
   Fire broke out close here, premises not in
      danger . . . . . . Deflagro
   Fire broke out in adjoining premises, ours in
      danger . . . . . . Deflexi
   Fire broke out in adjoining premises, ours safe Deformis
   Have had a *fire,* business as usual . . Defossus

Fix a meeting for — . . . . . . Degener
   Fix a meeting for any day convenient to — Deglubo
   What date is the meeting *fixed* for . . Degusto

Follow it up immediately . . . . Dehisco
   The remainder will *follow* by next mail . Deinceps

(Forget)—Do not *forget* — . . . . Delabor
   Do not *forget* to bring — . . . . Delacero
   Do not *forget* to come this evening . . Delego
   Do not *forget* to-night's engagement . . Deletrix
   Do not *forget* your appointment for to-day . Delinquo
   Do not *forget* your appointment for to-morrow Deliro
   Forgotten my bag, please keep until — . Deltoton
   Forgotten my books, please keep until — . Delubrum

Forgotten my box, please keep until — .	Demando
Forgotten my keys, please keep until — .	Demetor
Forgotten my letters, please keep until — .	Demigro
Forgotten my luggage, please keep until —.	Demitto
Forgotten my overcoat, please keep until —.	Demorior
Forgotten my papers, please keep until — .	Denarius
Forgotten my parcel, please keep until — .	Denascor
Forgotten my portmanteau, please keep until	Denique
Forgotten my purse, please keep until — .	Dens
Forgotten my things, please keep until — .	Dentale
Forgotten my umbrella, please keep until —	Dentitio
Forgotten my waterproof, please keep until —	Denudo
Forgotten my bag, please send immediately to —	Denumero
Forgotten my books, please send immediately to —	Depactus
Forgotten my box, please send immediately to —	Dependo
Forgotten my keys, please send immediately to —	Deplumis
Forgotten my letters, please send immediately to —	Depopulo
Forgotten my luggage, please send immediately to —	Deprimo
Forgotten my overcoat, please send immediately to —	Depso
Forgotten my papers, please send immediately to —	Depulsio
Forgotten my parcel, please send immediately to —	Depygis
Forgotten my portmanteau, please send immediately to —	Derosus
Forgotten my purse, please send immediately to—	Derumpo
Forgotten my things, please send immediately to —	Descendo

Forgotten my umbrella, please send immediately to — . . . . . Describo
Forgotten my waterproof, please send immediately to — . . . . . Desertor

Forward. (Refer to SEND, LETTERS.)
Forward any communications there may be waiting for me to — . . . . Desidia
Forward my luggage here . . . . Designo
Forward my luggage to — . . . Desisto
Forward my things here . . . . Desitum
Forward my things to — . . . . Despecto
How have my letters been *forwarded* . . Despicor
They shall be *forwarded* by next post . . Despolio
When will you *forward* . . . . Desubito
Will be *forwarded* immediately . . . Desuper

Freight cannot be obtained . . . . Deterius
Freight has been engaged . . . . Detorno
Cannot obtain *freight* enough . . . Devectus
What *freight* can you obtain . . . Devello

Funeral. (Refer to DEATHS.)

Get on quickly with — . . . . . Dextans
Get on quickly with work . . . . Diadema

Glad to hear from you . . . . . Diallage

Go as quickly as possible . . . . Diametros
Go as quickly as possible to — . . . Dianome

Goods. (Refer to ARRIVAL, INVOICES, ORDER, PATTERN, QUALITY, QUANTITY, SAMPLE, SEND.)
Goods arrived in bad condition . . . Diapente

Goods arrived in bad condition ; letter follows	Diatoni
Goods arrived safe	Diaulus
Goods arrived slightly damaged ; letter follows	Dicatura
Goods left by — . . . • . .	Dictator
Goods not arrived —	Diduco
Goods not arrived ; make enquiries . .	Diecula
Goods not arrived ; make enquiries. Am doing so here	Diffamo
Goods not arrived ; when did they leave .	Diffido
Goods not arrived ; when did they leave, and how	Diffuse
Goods on order waiting remittance . .	Digero
Have you received the *goods* . . .	Digestio
Have you received the *goods* sent by — .	Dignosco
Have you received the *goods* sent on the — .	Dijudico

Guarantee must be given	Dilamino
The quality must be *guaranteed* . . .	Dilapsus
Who will *guarantee* us	Dilorico
Will *guarantee* to the extent of — . .	Dilutium
Will you *guarantee* —	Dimadeo

Happy New Year	Dimidius
Happy New Year to all	Diminuo
Happy New Year to all at home . . .	Dimotus

Health. (Refer to Business, Expect, Medicine, Progress.)

Amputation is considered necessary . .	Dionysia
Amputation is considered unnecessary . .	Diota
Attack is considered serious . . .	Diphris
Attack is considered not serious . . .	Diploma
Attack is considered trifling . . .	Dipsacon
Has *changed* for the worse, but doctor gives hope	Dipsas

Has *changed* for the worse and doctor gives no hope	Diradio
Has *changed* for the worse and doctor gives no hope, come quickly	Directio
Has *changed* for the worse and doctor gives no hope, useless your coming . .	Direxi
Condition has changed slightly for better .	Discoquo
Condition is unchanged	Discretus
A *consulting physician* has been called in .	Discurro
A *consulting physician* is not required . .	Dispando
Continues to improve	Dispar
Is able to attend to *correspondence* . .	Dispenso
Is unable to attend to *correspondence* . .	Displodo
Doctor considers the crisis safely over . .	Dispudet
Doctor has ordered solid food . . .	Disquiro
Doctor is more hopeful to-day . . .	Dissero
Doctor states the illness to be — . .	Dissideo
Consult your *doctor*	Dissocio
Consult your *doctor* and let me know result .	Dissono
Have consulted *doctor* and he considers — .	Distraxi
Have consulted *doctor*, will post you particulars	Ditesco
Fever is increasing	Diva
Fever is subsiding	Divarico
Is able to *go out* for a drive . . .	Diverse
Is unable to *go out* for a drive . . .	Divisio
Is able to *go out* for a walk . . .	Diurnus
Is unable to *go out* for a walk . . .	Docilis
Illness commenced on the — . . .	Docte
The *improvement* is sustained . . .	Doctrina
The *improvement* is not sustained . .	Dogma
Invalid better; doctor recommends change of air	Dolatus
Invalid is now out of danger . . .	Dolenter
Invalid is now quite well	Dolium

Invalid is now quite well and will write you	Domator
Invalid recommended change of air by doctor, and as soon as convalescent will go to —	Domitrix
Invalid the same, doctor recommends change of air	Domo
Is able to undertake the *journey* . . .	Donarium
Is unable to undertake the *journey* . .	Donax
Can take no *nourishment*	Dormisco
Can take no *nourishment*, come if possible .	Drapeta
Operation has been performed successfully .	Draucus
Operation has been performed without success	Dryades
Patient has become unconscious . . .	Dubito
Patient is now quite conscious . . .	Ducatus
Patient is still unconscious . . .	Ducenti
Is able to be *removed* from bed . . .	Duellum
Is unable to be *removed* from bed . .	Dulcedo
Is able to be *removed* to another room . .	Dulcifera
Is unable to be *removed* to another room .	Dummodo
Passed a *sleepless* night, and is feverish this morning	Duodecim
Slept well, and has taken nourishment this morning	Duplex
Is rapidly regaining *strength* . . .	Duplico
Symptoms are not considered alarming. .	Duramen
Symptoms show great improvement . .	Dureta
Symptoms show no improvement. . .	Durities
Taken ill	Dysuria
Taken ill, cannot come	Fabacia
Taken ill, cannot come to-day . . .	Fabarius
Taken ill, cannot come this week. . .	Fabrica
Taken ill, cannot keep appointment . .	Fabrilis
Taken ill, cannot leave	Fabularis
Taken ill slightly, do not be alarmed . .	Fabulose
Taken ill slightly, letter by post. . .	Facesso
Taken ill slightly, no necessity for doctor .	Facetus

Taken ill slightly, no necessity for doctor, will write	Facilis
Taken ill slightly, will telegraph again .	Facticius
Taken ill slightly, will telegraph again if no better	Factito
Taken ill slightly, will telegraph again if worse	Facula
Taken ill slightly, unable to travel to-day .	Facultas
Taken ill slightly, unable to see you to-day	Fageus
Taken ill suddenly	Fagineus
Taken ill suddenly, come at once . .	Falarica
Taken ill suddenly, send doctor immediately.	Falcifer
Taken ill suddenly, send doctor and come yourself	Falco
Taken ill suddenly, and dangerously . .	Falere
Taken ill suddenly, and dangerously, come at once	Falernum
Taken ill with fever	Faliscus
Taken ill with fever, do not come . .	Fallacia
Taken ill with fever, let children remain .	Fallax
Telegraph health of —	Falsus
Telegraph health of *Baby*	Famiger
Telegraph health of *Baby*, shall I come .	Famosus
Telegraph health of *Brother* . . .	Famulor
Telegraph health of *Brother*, shall I come .	Fandus
Telegraph health of *Children* . . .	Fanum
Telegraph health of *Children*, shall I come .	Farcimen
Telegraph health of *Daughter* . . .	Farcio
Telegraph health of *Daughter*, shall I come	Farctura
Telegraph health of *Father* . . .	Farnus
Telegraph health of *Father*, shall I come .	Farrago
Telegraph health of *Grandfather* . .	Fartilis
Telegraph health of *Grandfather*, shall I come	Fascino

D

Telegraph health of *Grandmother* . .	Fasciola
Telegraph health of *Grandmother*, shall I come	Fastigo
Telegraph health of *Husband* . . .	Fastosus
Telegraph health of *Husband*, shall I come .	Fatifer
Telegraph health of *invalid* . . .	Fatisco
Telegraph health of *Mother* . . .	Fatuitas
Telegraph health of *Mother*, shall I come .	Fatum
Telegraph health of *Sister*	Favere
Telegraph health of *Sister*, shall I come .	Favilla
Telegraph health of *Son*	Februa
Telegraph health of *Son*, shall I come . .	Fecialis
Telegraph health of *Wife*	Fecundo
Telegraph health of *Wife*, shall I come .	Felineus
Telegraph health of *yourself* . . .	Felix
Telegraph health of *yourself*, shall I come .	Fellator
Telegraph health of *yourself*, am very anxious	Femella
Very much better	Femur
Very much better, do not come, will write .	Fenebris
Very much better, and improving fast .	Fenero
Very much better, and improving fast, do . not come	Fenestra
Wound is healing rapidly	Feralia
Wound is healing satisfactorily . . .	Fereola
Wound is not healing satisfactorily . .	Feretrum
Wound is not healing very rapidly . .	Ferio
Please *write* fully present condition of patient	Ferme
(High) How *high* can I go	Fermento
How *high* can you go	Ferocia
It is too *high*	Ferox
Hire if possible	Ferreus
Hire a conveyance to meet me at — . .	Ferrugo

Can you *hire*	Ferrum
Do not *hire*	Fervesco

Home by first train in morning . . .	Fervidus
Home by last train to-night . . .	Ferula
Home by first train on —	Fervor
Shall not be *home* this evening . . .	Festiva
Shall not be *home* this evening until —	Festuca

Hotels. (Refer to FORGET, FORWARD, LUG-GAGE.)

Shall *arrive* about — o'clock . . .	Fiber
Shall *arrive* by the mail train . . .	Fibra
Shall *arrive* by the steamer due on —. .	Fibratus
Shall *arrive* by the train due at — .	Ficedula
Reserve a *single bedroom* for me to-night .	Ficetum
Reserve comfortable *single bedroom* for me to-night	Ficosus
Reserve a *single bedroom* for me to-night, shall arrive at —	Fictilis
Reserve comfortable *single bedroom* for me to-night, shall arrive at — . . .	Fictio
Reserve a *single bedroom* for me to-night, shall require —	Ficulus
Reserve comfortable *single bedroom* for me to-night, shall require — . . .	Fidenter
Reserve a *single bedroom* for me to-morrow .	Fidicen
Reserve a comfortable *single bedroom* for me to-morrow	Fidicula
Reserve a *single bedroom* for me to-morrow, shall arrive at —	Fiducia
Reserve a comfortable *single bedroom* for me to-morrow, shall arrive at —. . .	Figlinus
Reserve a *single bedroom* for me, arriving on —	Figulina

D 2

Reserve a comfortable *single bedroom* for me,
arriving on — . . . . . **Filatim**
Reserve a *double bedroom* for me for to-night **Filictum**
Reserve a *double bedroom* for me, not too
high up, for to-night . . . . **Filiola**
Reserve a *double bedroom* for me on — . **Filiolus**
Reserve a *double bedroom* for me, not too
high up, on — . . . . . **Filius**
Reserve a *double-bedded room* for to-night . **Filum**
Reserve a *double-bedded room*, not too high
up, for to-night . . . . . **Fimbria**
Reserve a *double-bedded room* for — . . **Finitio**
Reserve a *double-bedded room*, not too high
up, for — . . . . . . **Firmamen**
Reserve a *single bedroom* and sitting-room
to-night, shall arrive — . . . **Firmitas**
Reserve a *single bedroom* and sitting-room
to-night, shall arrive at — . . . **Firmus**
Reserve a *single bedroom* and sitting-room
to-night, shall require — . . . **Fiscina**
Reserve a *single bedroom* and sitting-room
to-morrow . . . . . . **Fissilis**
. Reserve a *double bedroom* and sitting-room
to-night, shall arrive at — . . . **Fissio**
Reserve a *double bedroom* and sitting-room
to-night, shall require — . . . **Fissura**
Reserve a *double bedroom* and sitting-room
to-morrow . . . . . . **Fistula**
Reserve two *single bedrooms* for to-night,
shall arrive at — . . . . . **Flabilis**
Reserve two *double bedrooms* for to-night,
shall arrive at — . . . . . **Flabrum**
Reserve two *double-bedded rooms* for to-
night, shall arrive at — . . . **Flaccida**
Reserve a *single bedroom* and a *double bedroom* for to-night, shall arrive at — . **Flaccus**

Reserve a *double bedroom* and a *double-bedded room* for to-night, shall arrive at —   Flagello

Reserve two *single bedrooms* for — ' .   .   Flagito

Reserve two *double bedrooms* for — .   .   Flagro

Reserve two *double-bedded rooms* for — .   Flagrum

Reserve a *single bedroom* and a *double bedroom* for — .   .   .   .   .   .   Flamen

Reserve a *single bedroom* and a *double-bedded room* for — .   .   .   .   .   .   Flamma

Reserve a *double bedroom* and a *double-bedded room* for — .   .   .   .   Flammeus

Reserve two *single bedrooms* and a sitting-room to-night, shall arrive at — .   .   Flammula

Reserve two *double bedrooms* and a sitting-room to-night, shall arrive at — .   .   Flatus

Reserve two *double-bedded rooms* and a sitting-room to-night, shall arrive at —   Flaveo

Reserve a *single bedroom*, a *double bedroom*, and a sitting-room for to-night, shall arrive at — .   .   .   .   .   .   Flavesco

Reserve a *single bedroom*, a *double-bedded room*, and a sitting-room for to-night, shall arrive at — .   .   .   .   .   Flecto

Reserve a *double bedroom*, a *double-bedded room*, and a sitting-room for to-night, shall arrive at — .   .   .   .   Fletifer

Reserve two *single bedrooms* and a sitting-room for — .   .   .   .   .   .   Flexuose

Reserve two *double bedrooms* and a sitting-room for — .   .   .   .   .   .   Flexuosus

Reserve two *double-bedded rooms* and a sitting-room for — .   .   .   .   Flexura

Reserve a *single bedroom*, a *double bedroom*, and a sitting-room for — .   .   .   Flictus

Reserve a *single bedroom*, a *double-bedded room*, and a sitting-room for — .   .   Fligo

Reserve a *double bedroom*, a *double-bedded room*, and a sitting-room for — . . **Floralis**

Reserve three *single bedrooms*, shall arrive on — **Floreus**

Reserve three *double bedrooms*, shall arrive on — . . . . **Floridus**

Reserve three *double-bedded rooms*, shall arrive on —. . . . . . **Florifer**

Reserve one *single bedroom* and two *double bedrooms*, shall arrive on — . . . **Fluctus**

Reserve one *single bedroom*, one *double bedroom*, and one *double-bedded room*, shall arrive on —. . . . . . **Fluentum**

Reserve one *single bedroom* and two *double-bedded rooms*, shall arrive on — . . **Fluta**

Reserve two *double bedrooms* and one *double-bedded room*, shall arrive on — . . **Fluvidus**

Reserve one *double bedroom* and two *double-bedded rooms*, shall arrive on — . . **Fluxio**

Reserve three *single bedrooms* and a sitting-room, shall arrive on — . . . **Focale**

Reserve three *double bedrooms* and a sitting-room, shall arrive on — . . . **Focaneus**

Reserve three *double-bedded rooms* and a sitting-room, shall arrive on — . . **Focillo**

Reserve one *single bedroom*, two *double bedrooms*, and a sitting-room, shall arrive on — **Fodico**

Reserve one *single bedroom*, one *double bedroom*, one *double-bedded room*, and a sitting-room, shall arrive on — . . **Foliosus**

Reserve one *single bedroom*, two *double-bedded rooms*, and a sitting-room, shall arrive on —. . . . . . **Folium**

Reserve two *double bedrooms*, one *double-bedded room*, and a sitting-room, shall arrive on —. . . . . . **Follitim**

Reserve one *double bedroom*, two *double-bedded rooms*, and a sitting-room, shall arrive on —	Fomenta
Reserve also one *maid-servants' bedroom* .	Forabilis
Reserve also two *maid-servants' bedrooms* .	Foramen
Reserve also one *man-servants' bedroom* .	Forceps
Reserve also two *men-servants' bedrooms* .	Forem
Reserve *beds* for two gentlemen to-night .	Forensis
Reserve *beds* for three gentlemen to-night .	Forfex
Reserve *beds* for four gentlemen to-night .	Forica
Reserve *beds* for two gentlemen to-morrow .	Igneus
Reserve *beds* for three gentlemen to-morrow	Ignigena
Reserve *beds* for four gentlemen to-morrow .	Ignipes
Reserve *beds* for two gentlemen for — .	Ignoro
Reserve *beds* for three gentlemen for — .	Ilex
Reserve *beds* for four gentlemen for — .	Ilicit
Have a *carriage* with single horse ready for me at — o'clock	Ilicitum
Have a *carriage* with single horse at the station to meet train arriving — . .	Iligneus
Have a *carriage* with two horses ready for me at — o'clock	Iliosus
Have a *carriage* with two horses at the station to meet train arriving — . .	Illabor
Have a *fly* ready for me at — o'clock . .	Illac
Have a *fly* at the station to meet train arriving —	Illaqueo
Have an *omnibus* at the station to meet train arriving —	Illapsa
Have *dinner* ready this evening punctually at — o'clock	Illatro
Require *dinner* this evening in private room, for — persons	Illicio
Reserve seats at the Table d'hôte *dinner* for — persons	Illitus

Have a *fire* in the bedroom	Illuceo
Have *fires* in the bedroom and sitting-room .	Illumino
— *found* and forwarded as instructed . .	Imaginor
What *Hotel* do you recommend in — . .	Imago
— when did he *leave* your Hotel . .	Imberbis
— *left* in Hotel, send to me at — . .	Imbibo
— *left* in Hotel, send to me here . .	Imbrex
— *left* in my room, send to me at — .	Imbrifer
— *left* in my room, send to me here . .	Imbuo
— *left* this Hotel on the — . . .	Imitamen
— *left* this Hotel to go to — . . .	Imitatio
— has not yet *left* this Hotel . . .	Immadeo
Have sent articles *left* in your room as instructed	Immemor
Send articles *left* in my room to — . .	Immensus
Send articles *left* in my room to this address	Immergo
Have sent luggage *left* in your room, as instructed	Immigro
Send luggage *left* in my room to — . .	Imminuo
Send luggage *left* in my room to this address	Immisceo
Any *letters* or other communication arriving for me send to —	Immixtus
Any *letters* or *telegrams* arriving for me keep until I arrive	Immodice
Any *letters* or *telegrams* arriving for me redirect them to—	Immolo
Any *letters* or *telegrams* arriving for me redirect them to this address . . .	Immorior
Some *letters* lying here for you, what shall be done with them	Immotus
Some *letters* and *telegrams* lying here for you, what shall be done with them .	Immugio
Some *telegrams* lying here for you, what shall be done with them	Impages
— *missing* on arrival here, can you trace it	Impasco

— *missing* on arrival, can you trace it? if so
   forward to me — .   .   .   .   Impavide
— *missing* on arrival here, have search made
   for it and telegraph me result   . . Impendeo
Can find no trace of the *missing* articles . Impensus
Rooms reserved as requested   .   .   . Imperium
Rooms reserved as requested and your orders
   will be attended to   .   .   .   . Impete
Cannot reserve the *rooms* asked for. Already
   full .   .   .   . Impetigo
Please keep my *rooms* ; cannot arrive until — Impexus
Shall not require *rooms* ordered .   .   . Impietas
Shall require *supper* on arrival .   .   . Impiger

If you have not already done so.   .   .   . Impilia
   If you have not already done so, do not send Impingo
   If you have not already done so, send at once Impius

Illness.   (Refer to HEALTH.)

(Include)—Does it *include* — .   .   .   . Implecto
   It does not *include* —   .   .   .   . Imploro
   Were *included* in estimates   . .   . Implumbo
   Were not *included* in estimates .   .   . Implumis
   Are the terms *inclusive*   .   .   .   . Impluo
   Terms are *inclusive* .   .   .   .   . Impolitus
   Terms are not *inclusive*   .   .   .   . Impono

(Inconvenience)—Can it be done without *inconvenience* .   .   .   .   .   . Impos
   It can be done without *inconvenience* . . Impositio

(Inform)—Can you *inform* me   .   .   . Imprecor
   Information has been received   .   .   . Imprimis
   From what source is the *information* . . Improbe
   More *information* is necessary .   .   . Impropero

No *information* has been received   . . Impuber
This is for your private *information* . . Impugno
Have you *informed* — . . . Impulsio

Inquire. (See ENQUIRE.)

.**Insurance** has been effected . . . Impunis
  Insure immediately . . . . Imulus
  Cannot *insure* under — . . . Inambulo

(Interfere)—Decline to *interfere* . . . Inanimus
  Will not *interfere* . . . . Inanio

Interview will be necessary . . . Inaniter
  Interview will not be necessary . . Inaratus

(Introduce)—Can you *introduce* me to — . Inaresco
  Introductions are sent by post . . Inaudax

Investigate into the matter immediately . . Inauguro
  Result of *investigations* not satisfactory . Incanto
  Result of *investigations* satisfactory . . Incassum

Invitations. (Refer to DINNER ENGAGEMENTS,
   LUNCH ENGAGEMENTS, THEATRE.)
  Accept your invitation with pleasure . . Incautus
  Regret I cannot *accept* your invitation . Incedo
  Shall I *accept* the invitation for — . . Incensus
  Shall you *accept* the invitation for — . . Inceptor
  Bring with you to-day your friend . . Incilis
  Bring your brother with you . . Incipio
  Bring your daughter with you . . Incisura
  Bring your father with you . . Incitate
  Bring your husband with you . . Inclinis
  Bring your mother with you . . Includo '
  Bring your sister with you . . Inclytus

Bring your son with you	Incoctus
Bring your wife with you	Incogito
Received invitation for —	Incoquo
Received invitation, shall I accept it, for —	Increbro
Received invitation to party to-night, at —.	Incrusto
Received invitation to party to-night. Can you come to —	Incubito
Reply, accepting the invitation . . .	Incubo
Reply, declining the invitation . . .	Incumbo
Has not *replied* to the invitation. . .	Incursio
Have not yet *replied* to the invitation. .	Incurvus
If you can arrange to *stay* to-night we can give you a bed	Indemnis
Shall be very glad if you can come and *stay* with us to-night	Indeptus
Shall be very glad if you can come and *stay* with us until —	Indicium
When can you come and *stay* with us for a few days	Indidem
Invoices have arrived	Indigena
Invoices have not arrived	Indigne
Invoices will be sent on —	Indoles
It is so	Indormio
It is not so	Inductio
It will do in default of better . . .	Inducula
It will do quite well	Indulgeo
It will not do	Indusium
(Judge)—You can best *judge* on the spot. .	Indutus
Keep possession until you hear from — .	Induxi
Will not *keep* it longer than — . . .	Inemptus
Know nothing whatever of the matter . . .	Ineptia

Known, and in good position   .   .   . Inequito
Is not *known* here   .   .   .   .   . Inertia

Leave all matters in your hands   .   .   . Infabre
Leave everything just as it is   .   .   . Infacete
Leave has been given   .   .   .   . Infamis
Leave has been refused   .   .   .   . Infandus
Leave immediately on receipt of this . . Infantia
Cannot *leave* here at present   .   .   . Infarcio

Leaving.   (Refer to Departure.)
Leaving on the 1st by the Steamship — . Infatuo
Leaving on the 2nd by the Steamship — . Infector
Leaving on the 3rd by the Steamship — . Infelix
Leaving on the 4th by the Steamship — . Inferveo
Leaving on the 5th by the Steamship — . Infestus
Leaving on the 6th by the Steamship — . Infibulo
Leaving on the 7th by the Steamship — . Infindo
Leaving on the 8th by the Steamship — . Infirmo
Leaving on the 9th by the Steamship — . Inflammo
Leaving on the 10th by the Steamship — . Inflatus
Leaving on the 11th by the Steamship — . Inflexio
Leaving on the 12th by the Steamship — . Infloreo
Leaving on the 13th by the Steamship — . Informis
Leaving on the 14th by the Steamship — . Infossus
Leaving on the 15th by the Steamship — . Infra
Leaving on the 16th by the Steamship — . Infundo
Leaving on the 17th by the Steamship — . Infusio
Leaving on the 18th by the Steamship — . Ingemino
Leaving on the 19th by the Steamship — . Ingemo
Leaving on the 20th by the Steamship — . Ingenium
Leaving on the 21st by the Steamship — . Ingenue
Leaving on the 22nd by the Steamship — . Ingestum
Leaving on the 23rd by the Steamship — . Ingrate
Leaving on the 24th by the Steamship — . Ingravo
Leaving on the 25th by the Steamship — . Ingressio

Leaving on the 26th by the Steamship — .	Ingruo
Leaving on the 27th by the Steamship — .	Inhabito
Leaving on the 28th by the Steamship — .	Inhorreo
Leaving on the 29th by the Steamship — .	Inhumo
Leaving on the 30th by the Steamship — .	Inimico
Leaving on the 31st by the Steamship — .	Inique

Legal.　(Refer to POSTPONE, POWER.)

Appeal has been allowed, costs in the cause .	Initio
Appeal has been allowed, with costs　.　.	Injectio
Appeal has been dismissed, costs in the cause	Injungo
Appeal has been dismissed, with costs .　.	Injura
Attorney for the defendant is — .　.　.	Injussus
Attorney for the plaintiff is — .　.　.	Injuste
Attorney on the other side is — .　.　.	Innato
Who is the *attorney* for — .　.　.　.	Innitor
Who is the *attorney* for the defendant　.	Innocens
Who is the *attorney* for the plaintiff .　.	Innovo
Who is the *attorney* on the other side .　.	Innoxius
Bail has been accepted　.　.　.　.	Innuptus
Bail has been refused　.　.　.　.	Innutrio
Bail is required for — .　.　.　.	Inocco
Counsel for the *defendant* has commenced his cross-examination .　.　.　.	Inoculo
Counsel for the defendant has finished his cross-examination　.　.　.　.	Inodorus
Counsel for the defendant has commenced his speech .　.　.　.　.　.	Inopaco
Counsel for the defendant has finished his speech　.　.　.　.　.　.	Inopinus
Counsel for the defendant has commenced calling his witnesses .　.　.　.	Inoratus
Counsel for the defendant has finished calling his witnesses　.　.　.　.　.	Inornate
Counsel for the *plaintiff* has commenced his cross-examination.　.　.　.　.	Inquam

Counsel for the plaintiff has finished his cross-examination	Inquiro
Counsel for the plaintiff has commenced his speech	Insanio
Counsel for the plaintiff has finished his speech	Inscendo
Counsel for the plaintiff has commenced calling his witnesses	Insciens
Counsel for the plaintiff has finished calling his witnesses	Inscribo
Opinion of counsel is against them . .	Insculpo
Opinion of *counsel* is against us . . .	Inseco
Opinion of *counsel* is against you . .	Insequor
Opinion of *counsel* is in our favour . .	Insertim
Opinion of *counsel* is in their favour . .	Inservio
Opinion of *counsel* is in your favour . .	Insidior
Retain as *counsel* on my behalf — . .	Insido
Defendant has retained as his *counsel* — .	Insignis
Have retained as *counsel* on your behalf — .	Insimulo
Plaintiff has retained as his *counsel* — .	Insisto
Case is in the list for *hearing* to-morrow .	Insolens
Case is not in the list for *hearing* to-morrow	Insolida
When will the case be in the list for *hearing*	·Inspergo
Hearing of the case commenced to-day, adjourned until —	Inspicio
Hearing of the case commenced to-day, adjourned until to-morrow . . .	Insterno
Hearing of the case concluded, judgment reserved	Instillo
Hearing of the case concluded, jury have retired	Instruo
Hearing of the case will be concluded to-day	Insuasum
Hearing of the case will be concluded to-morrow	Insula

Hearing of the case will be concluded this week	Insurgo
Hearing of the case will last for several days	Intectus
Judge has commenced his summing-up .	Integro
Judge has concluded his summing-up . .	Maceria
Judge has concluded his summing-up. It was in favour of defendant . . .	Macies
Judge has concluded his summing-up. It was in favour of neither party . .	Mactator
Judge has concluded his summing-up. It was in favour of plaintiff . . .	Madefeci
Jury are still absent	Madulsa
Jury could not agree, and have been discharged	Magnus
Subpœna as a witness on our side — . .	Majestas
Subpœna has been served on — . . .	Malache
Subpœna has not been served, as we cannot find —	Maledice
Summons has been issued, returnable on —	Maligne
Summons heard to-day. Adjourned for a week	Manceps
Summons heard to-day. Adjourned into court	Manedum
Summons heard to-day. Adjourned until —	Manesis
Summons heard to-day. Chief clerk reserves decision	Manifeste
Summons heard to-day. Master reserves decision	Manliana
Summons heard to-day. Order made as asked	Mansito
Summons heard to-day. Order refused .	Mansuete
Summons heard to-day. Order refused, with costs	Marceo

Summons heard to-day. Referred to judge in chambers	Margaris
Verdict against us	Margino
Verdict for —	Marinus
Verdict for the defendant, with costs . .	Masculus
Verdict for the defendant, without costs .	Maspetum
Verdict for the plaintiff, damages — . .	Mastigia
Verdict for the plaintiff, with costs . .	Mastruca
Verdict for the plaintiff, without costs .	Matellio
Verdict in our favour	Mater
Verdict of guilty. Sentence — . .	Maternus
Verdict of guilty. Sentence deferred . .	Matralia
Verdict of manslaughter. Sentence — .	Matrimus
Verdict of manslaughter. Sentence deferred	Mature
Verdict of not guilty	Maxilla
Verdict of not proven	Maxime
Verdict of temporary insanity . . .	Meatus
Verdict of wilful murder	Mecastor
Witnesses must be in attendance on — .	Meconis
Your attendance in court as *witness* is required —	Medianus
Writ has been issued as requested . .	Medicina
Writ has been served on all the defendants .	Medicus
Writ has been served on the defendant .	Meditor
Writ has not yet been served on the defendant	Medium
Have obtained order for substituted service of *writ*	Megalium
Length is —	Melampus
What is the *length* of —	Melania
Let on the terms agreed the — . . .	Melitites
Let on the terms named the — . . .	Mellifer

Letter. (Refer to ACKNOWLEDGE, EXPECTING,
 FORWARD, HOTELS, POST, REPLY, WAIT,
 WRITING, WROTE.)

Letter last received was dated — . .	**Melligo**
Been obliged to leave before the arrival of your *letter.* Telegraph to — . .	**Mellilla**
Been obliged to leave before the arrival of your *letter.* Write again to — . .	**Mellinia**
Do not forward any more *letters.* Keep them until my return	**Mellis**
Forward all *letters* until further instructions to —	**Mellitorum**
Forward my *letters* to-day to Poste Restante —	**Mellitum**
Forward my *letters* to-day and to-morrow to Poste Restante —	**Mello**
Forward my *letters* until further instructions to Poste Restante —	**Mellorum**
Forward my *letters* to-day to Poste Restante here	**Mellum**
Forward my *letters* to-day and to-morrow to Poste Restante here	**Melofolia**
Forward my *letters* until further instructions to Poste Restante here	**Melofoliam**
Forward my *letters* to —	**Melomeli**
Have received *letter* from — . . .	**Melopepo**
Have not received *letter* from — . .	**Membrana**
Have not received your *letter* . . .	**Memento**
Have not received any *letter* from you since —	**Meminisse**
Have not received any *letter* from you. Telegraph at once to — . . .	**Memoris**
Have not received any *letter* from you. Write at once to —	**Memoratus**
Have you received my *letter* . . .	**Memoro**
Have you received my *letter* with enclosures	**Mendacis**

E

Have you received my *letter;* no reply received, telegraph	Mendose
Private *letter* posted you to-day . .	Mensio
Private *letter* posted to-day to — . .	Mensula
Private *letter* posted you yesterday . .	Mentagra
Private *letter* posted yesterday to — . .	Mentum
Private *letter* not received	Mercator
Private *letter* received	Merenda
Received your *letter*	Mergo
Received your *letter* and enclosures . .	Meridies
Received your *letter* and forwarded it as requested	Merulam
Received your *letter,* and instructions noted	Mespilum
Received your *letter* and returned it to you .	Metallum
Received your *letter* and telegram . .	Metaphora
Received your *letter* and telegram ; having attention	Metatio
Received your *letter* and will attend to it immediately	Metopion
Received your *letter* and will attend to it as soon as possible	Metreta
Received your *letter;* attending to contents	Metuo
Received your *letter;* attended to contents and writing	Migratio
Received your *letter,* but no enclosures .	Militia
Received your *letter;* cannot do as you wish	Milvago
Received your *letter;* cannot reply until —	Milvinus
Received your *letter* containing cheque .	Minister
Received your *letter* containing P.O.O. .	Minium
Received your *letter;* do not understand — .	Minor
Received your *letter;* do not understand first portion	Minutal
Received your *letter ;* do not understand latter portion	Mirmillo
Received your *letter ;* have already sent .	Miseria
Received your *letter ;* have replied fully .	Mithrax

Received your *letter;* order will be attended to	Mitra
Received your *letter;* orders will be attended to at once	Mixtim
Received your *letter;* too busy to attend to matter	Mixtura
Received your *letter;* too unwell to write .	Mobilior
Received your *letter;* too unwell to attend to matter	Modestia
Received your *letter;* will do as you wish .	Modius
Received your *letter;* will post you particulars of —	Modulate
What is the date of the last *letter* . .	Molimen

(Litigation)—Do your utmost to avoid *litigation* . . . . Molitura

Lodgings. (Refer to APARTMENTS.)

Lose no time in the matter	Mollesca
Do not *lose* an instant	Momentum
Do not *lose* an instant, but come at once .	Monedula

Loss. (Refer to FORGET, HOTELS.)

Luggage. (Refer to FORGET, FORWARD, HOTELS.)

Luggage has been left behind; enquire at the station	Monitor
Leave your *luggage* at the cloakroom to be called for	Monstro
Send my *luggage* on to —	Montuosus
Send a conveyance to take the *luggage* .	Moralis
Will send a man to take charge of the *luggage*	Morbidus

Lunch. (Refer to INVITATIONS.)

Will *lunch* with you — . . . . Morbonia

E 2

Will *lunch* with you to-day . . .	Mordax
Will *lunch* with you to-day, and wait your arrival at —	Morigero
Will *lunch* with you to-day, and will call at —	Mormyr
Will *lunch* with you to-morrow . . .	Morose
Will *lunch* with you to-morrow, and wait your arrival	Morpheus
Will *lunch* with you and call at — . .	Motus
Will *lunch* with you on Monday . .	Mucidus
Will *lunch* with you on Tuesday . .	Muginor
Will *lunch* with you on Wednesday . .	Mugitus
Will *lunch* with you on Thursday . .	Mulctrum
Will *lunch* with you on Friday . . .	Multifer
Will *lunch* with you on Saturday . .	Munditer
Will *lunch* with you on Sunday . . .	Municeps
Will you *lunch* with me — . . .	Munifice
Will you *lunch* with me to-day at — . .	Murcidus
Will you *lunch* with me to-day at the Club at —	Muraena
Will you *lunch* with me here at — . .	Musca
Will you *lunch* with me to-day; will call for you at —	Muscosus
Will you *lunch* with me to-day; call for me at —	Museus
Will you *lunch* with me to-morrow at — .	Mussito
Will you *lunch* with me to-morrow at the Club	Mustace
Will you *lunch* with me to-morrow here at —	Mutatio
Will you *lunch* with me to-morrow; will call for you at —	Myiagros
Will you *lunch* with me to-morrow; call for me at —	Myoparo
Will you *lunch* with me on Monday at — .	Myrapium
Will you *lunch* with me on Tuesday at — .	Myropola
Will you *lunch* with me on Wednesday at—	Myrrha

Will you *lunch* with me on Thursday at . Myscus
Will you *lunch* with me on Friday at — . Myxa
Will you *lunch* with me on Saturday at — . Nablium
Will you *lunch* with me on Sunday at — . Nactus

Machine has arrived . . . . . Nanus
Machine has arrived, and works well . . Nape
Machine has arrived, but works badly. . Naphtha
Machinery out of order, delay will be great . Napina

Many happy returns of the day . . . Nardinus
How *many* did you send . . . . Nardum
How *many* do you want . . . . Narrator

Market fully supplied . . . . . Narratus
Market fully supplied, prospects bad . . Narro
Market glutted, prospects very bad . . Narthex
Market very flat . . . . . Nascor

Marriage has been arranged between — . . Nasum
Marriage is announced of — . . . Nasutus
Marriage postponed in consequence of — . Natalis
Marriage postponed indefinitely . . . Natatio
Marriage postponed, particulars by letter . Natator
Marriage postponed until — . . . Nativus
Marriage takes place at —. . . . Natta
Marriage takes place on —. . . . Naulium
Marriage takes place on Monday. . . Nauplius
Marriage takes place on Tuesday. . . Nausea
Marriage takes place on Wednesday . . Nauseola
Marriage takes place on Thursday . . Nauticus
Marriage takes place on Friday . . . Naviger
Marriage takes place on Saturday . . Navigium
Marriage takes place on Sunday. . . Navigo
Marriage takes place January . . . Navorum
Marriage takes place February . • . Nebritis

Marriage takes place March . . .	Necesse
Marriage takes place April. . . .	Necessum
Marriage takes place May	Necnon
Marriage takes place June	Nectarea
Marriage takes place July	Necubi
Marriage takes place August . . .	Necunde
Marriage takes place September . . .	Nefandus
Marriage takes place October . . .	Nefarie
Marriage takes place November . . .	Nefarius
Marriage takes place December . . .	Nefas .
Marriage will not take place . . .	Nefastus
Marriage will not take place; particulars by letter	Negantia
Married to-day	Negatio
Married yesterday	Negito
Medicine not arrived	Negligo
Medicine not arrived, send at once . .	Negotium
Medicine wanted immediately . . .	Nepeta
Meet. (Refer to APPOINTMENTS.)	
Meet me at —	Nepos
Meet me at 1 o'clock at —. . . .	Neptis
Meet me at 1.30 at —	Nequam
Meet me at 2 at —	Neque
Meet me at 2.30 at —	Nequid
Meet me at 3 at —	Nequities
Meet me at 3.30 at —	Nervose
Meet me at 4 at —	Nervosus
Meet me at 4.30 at —	Nervulus
Meet me at 5 at —	Nescio
Meet me at 5.30 at —	Nescius
Meet me at 6 at —	Neuradis
Meet me at 6.30 at —	Neuricus
Meet me at 7 at —	Neutro
Meet me at 7.30 at —	Nexilis

Meet me at 8 at —	Nexum
Meet me at 8.30 at —	Nictatio
Meet me at 9 at —	Nicto
Meet me at 9.30 at —	Nidorem
Meet me at 10 at —	Nidulor
Meet me at 10.30 at —	Nidulus
Meet me at 11 at —	Nigellus
Meet me at 11.30 at —	Nigra
Meet me at 12 at —	Nigrina
Meet me at 12.30 at —	Nigresco
Meet me on Monday at —	Nigro
Meet me on Tuesday at —	Nihil
Meet me on Wednesday at — . . .	Nihildum
Meet me on Thursday at — . . .	Nilios
Meet me on Friday at —	Nilum
Meet me on Saturday at — . . .	Nimbifer
Meet me on Sunday at —	Nimbosus
Meet me this morning at — . . .	Nimbus
Meet me this afternoon at — . . .	Nimietas
Meet me this evening at — . . .	Nimio
Meet me to-night at —	Nimirum
Meet me to-morrow morning at — . .	Niteo
Meet me to-morrow evening — . . .	Nitescat
Meet me to-morrow afternoon at — . .	Nitido
Can you *meet* me at —	Nitidus
Can you *meet* me as suggested . . .	Nitraria
Can you *meet* me to-day	Nitratus
Can you *meet* me this evening . . .	Nitrum
Cannot *meet* you as arranged . . .	Nivarius
Cannot *meet* you as arranged, will explain later	Nivatus
Cannot *meet* you until —	Niveus
Meeting postponed	Nobilis
Meeting postponed until — . . .	Nobilito
Meeting takes place on — . . .	Nobis
Meeting takes place to-morrow . . .	Nocenter

Cannot attend the *meeting* of .the Board . Nocivus
Cannot attend the *meeting* of the Committee Noctifer

Met with. (Refer to ACCIDENT.)

Military. (Refer to HEALTH.)
   Consult the colonel and let me know result . Noctua
   Will you *exchange* with me   .   .   . Nodor
   Will you *exchange* with me for —   .   . Nodosus
   Will you sanction *exchange* with —   .   . Nominito
   Can my leave be *extended* to — .   .   . Nomino
   Extension of leave cannot be sanctioned . Nomus
   Extension of leave cannot be sanctioned, you
     must return at once   .   .   .   . Nonageni
   Extension of leave required   .   .   . Nonagies
   Extension of leave required on legal affairs
     until — .   .   .   .   . Nonanus
   Extension of leave required on urgent family
     affairs until — .   .   .   . Nonassis
   Extension of leave required on urgent pri-
     vate affairs until — .   .   . Nondum
   Extension of leave required, please sanction
     until — .   .   .   .   . Nonnemo
   Extension of leave required until — . Nonnihil
   Extension of leave sanctioned   .   . Norma
   Extension of leave sanctioned until — . Normalis
   Is my *extension* of leave sanctioned . . Noscito
   Furlough to all officers on leave has been
     cancelled, and they are ordered to rejoin
     their regiments by — .   .   . Nosco
   Furlough to all officers on leave has been
     cancelled, and they are ordered to rejoin
     at once   .   .   .   .   . Nossem
   The prohibition of *furlough* to officers on
     leave has been cancelled   .   .   . Noster
   You are *gazetted*   .   .   .   . Nostras

You are *gazetted* —	Notans
You are *gazetted* Lieutenant-Colonel . .	Notarius
You are *gazetted* Major	Notesco
You are *gazetted* Captain	Nothus
You are *gazetted* to a Company . . .	Novacula
You are *gazetted* to a Troop . . .	Novalis
— was *killed* in the engagement at — .	Novatrix
— was *killed* in the last engagement . .	Nove
Ordered home on sick leave . . .	Novello
The *Regiment* has been ordered to — . .	Noveni
The *Regiment* has been ordered to Canada .	Noverca
The *Regiment* has been ordered to Cyprus .	Novicius
The *Regiment* has been ordered to Egypt .	Novum
The *Regiment* has been ordered to England .	Noxia
The *Regiment* has been ordered to Gibraltar.	Noxiosus
The *Regiment* has been ordered to India .	Noxius
The *Regiment* has been ordered to Ireland .	Nubecula
The *Regiment* has been ordered to Malta .	Nuberum
The *Regiment* has been ordered to South Africa	Nubifer
The *Regiment* has been ordered to West Indies	Nubigena
The *Regiment* suffered little in the last engagement	Nubila
The *Regiment* suffered severely in the last engagement.	Nubilans
The *Regiment* took part in the battle at — .	Nucearum
The *Regiment* took part in the skirmish at—	Nucetum
Is *report* correct here that Regiment is ordered —	Nugator
Is *report* correct here that Regiment is ordered home	Nugax
Is *report* correct here that Regiment is ordered to Canada	Nullus
Is *report* correct here that Regiment is ordered to Cyprus	Numella

Is *report* correct here that Regiment is
ordered to Egypt . . . . . Numero
Is *report* correct here that Regiment is
ordered to Gibraltar . . . . Numisma
Is *report* correct here that Regiment is
ordered to India . . . . . Nummatus
Is *report* correct here that Regiment is .
ordered to Ireland . . . . Numne
Is *report* correct here that Regiment is
ordered to Malta . . . . . Numquis
Is *report* correct here that Regiment is
ordered to South Africa . . . Nunccine
Is *report* correct here that Regiment is
ordered to West Indies . . . . Nuncubi
War has been declared between — . . Nundinum
War is expected to be declared between — . Nunquam
May I *withdraw* papers . . . . Nunquid
You may *withdraw* papers . . . . Nuntio
You may not *withdraw* papers . . . Nuper
Was *wounded* severely . . . . Nuperus.
Was *wounded* slightly . . . . Nupta
Was not *wounded* nor hurt . . . . Nupturus

Miss the Train. (Refer to TRAIN.)

Missing. (Refer to FORGET, HOTELS, LUGGAGE.)

Mistake has been discovered and rectified. . Nuribus
    Mistake has not been made on this side . Nutamen
    Correct the *mistake* without delay . . Nutrico
    Have you discovered the *mistake*. . . Nutrimen

Money. (Refer to ACCOUNT, CHEQUE, PLACE,
    REMITTANCE.)
    Money is nearly exhausted. When will more
    be provided . . . . . . Nutritus

Money is very plentiful in the market . .	Nutrix
Money is very scarce in the market · . .	Nycteris
Money will be forthcoming when required .	Nympha
No *money*, cannot leave until debts paid, remit —	Obambulo
No *money*, in great difficulties, remit — .	Obarmo
No *money*, send at once £5	Obba
No *money*, send at once £10 . . .	Obcalleo
No *money*, send at once £20 . . .	Obdormio
No *money*, send at once £25 . . .	Obduco
No *money*, send at once £50 . . .	Obductio
No *money*, send at once £75 . . .	Obedio
No *money*, send at once £100 . . .	Obeliscus
No *money*, send at once £150 . . .	Oberro
No *money*, send at once £200 . . .	Obesitas
No *money*, send at once £250 . . .	Obfero
No *money*, pay passage at Agent's and telegraph	Obgannio
No *money*, telegraph some through —. .	Obiratus
No *money* to pay bills	Obitus
No *money* to pay bills before leaving, remit—	Objaceo
No *money* to pay passage, remit — .	Objectus
No *money* to pay passage, remit by telegraph	Oblatro
No *money* to pay wages, remit quickly. .	Oblatus
No *money*, very ill, and want to come home .	Oblique
Name you ask for is —	Oblisus
Do not know the *name* of — . . .	Oblivio
Send *name* of —	Oblivium
(Nature)—What is the *nature* of the communication from —	Oblocutor
Negotiations are pending	Oblongus
Negotiations are suspended temporarily .	Obloquor
Have broken off *negotiations* . . .	Obluctor

(Net)—Is it *net*	Obmolio
It is *net*	Obmotus
Quotation is *net*	Obmoveo
None is to be got	Obnitor
· **Note** my address at foot	Obnixe
Note my address at foot for the present	Obnuntio
Notice must immediately be given to .	Oboleo
Have given *notice* to — . . .	Oboritor
Have received *notice* from — . .	Obrigeo
Number you inquire for is —. . .	Obrodo
What is the *number* of — . . .	Obrussa
Object very strongly	Obsaturo
Have no *objection*	Obsecro
Have you any *objection* . . .	Obsepio
Obtain as much as possible . . .	Obsequor
Obtained what was wanted . .	Obsessio
Can you *obtain* —	Obsidium
It cannot be *obtained*	Obsolete
Offer is accepted	Obsono
Offer is refused	Obstipus
Offer was made too late . . .	Obstiti
Can you *offer* more on same terms .	Obsto
Cannot *offer* more	Obstrepo
Make an *offer* of —	Obstupeo
Opportunity has gone by . . .	Obsutus
Opportunity has not arisen . . .	Obtectus
Wait for a better *opportunity* . .	Obtero

Option will be given for — . . . . Obtestor
Option will be given until — . . . Obtorpeo
Cannot give *option* . . . . . Obtortus

Order. (Refer to CANCEL, GOODS.)
Order cannot be executed . . . . Obtrudo
Order cannot be executed until — . . . Obtrunco
Order executed before your telegram arrived . . . . Obumbro
Order is already executed . . . . Obuncus
All your *orders* have been executed . . Obvagio
Am without *orders* . . . . . Obversor
Cannot accept your *order* for — . . . Obverto
Have received your *order*. Goods will be despatched to-day . . . . Obvertunto
Have received your *order*. Goods will be despatched to-morrow . . . . Obvigilat
Have received your *order*. Goods will be despatched this week . . . . Obvigilavi
Have received your *order*. Goods will be despatched next week . . . . Obvigilo
Have received your *order*. Goods will be despatched in — weeks . . . Obviam
Have you executed our last *order* . . Obvius
Is not according to *order* . . . . Obvolvo
Reply by post when you will despatch our *order* of — . . . . . . Obvolvunt
Reply by telegram when you will despatch our *order* of — . . . . . Occallatus
This *order* is in addition to previous . . Occalleo
This *order* is in substitution of previous . Occanere
This *order* must leave you on — . . Occano
Wait cash before executing *order* . . Occasio
When is the earliest you can deliver *order* . Occasiones
Will accept your *order* for — . . . Occasus
Your *order* is being executed . . . Occator

(Owing)—What is *owing* from — .   .   . Occento
What is *owing* to — .   .   .   .   . Occentus

(Paid)—Has been *paid* .   .   .   .   . Occiduus
Has it been *paid*   .   .   .   . Occiput
Has not been *paid* .   .   .   . Occisor
Have you *paid* .   .   .   .   . Occludo
Must not be *paid*   .   .   .   . Occulco
Will be *paid* .   .   .   .   . Occurro
Will not be *paid*   .   .   .   . Oceanus

Parcel. (Refer to SEND.)
Parcel has been forwarded .   .   .   . Ocellus
Parcel is waiting remittance   .   .   . Ochra
Parcel must be forwarded by passenger
  train .   .   .   . Ocimum
Parcel received all right .   .   .   . Ocrea
Have not received the *parcel*   .   .   . Octans
Have you received the *parcel*   .   .   . Octipes
How was the *parcel* forwarded .   .   . Octogeni
When was the *parcel* forwarded .   .   . Octuplus

Passage. (Refer to MONEY, SHIP, WEATHER.)
Passage paid here, call for particulars and
  tickets at — .   .   .   .   . Octussis
Passage paid here, telegraph departure, call
  for particulars at — .   .   .   . Odeum
Take *passage* by the — .   .   .   . Odi
Take *passage.* by the —, and telegraph
  departure .   .   .   .   .   . Odiosum
Take *passage*, and come at once by the — . Odor
Take *passage*, and come at once, telegraph
  departure .   .   .   .   .   . Odoratio

Patterns are suitable .   .   .   .   . Odoratus
Patterns are suitable, but material too cheap Odyssea

Patterns are suitable, but material too dear .	Ofella
Patterns are not suitable	Offatim
Can supply that *pattern* in — . . .	Offectus
Have you more of the same *pattern* in stock	Officina
Have you much of the same *pattern* in stock	Officiose
How soon can you supply more of the same *pattern*	Officium
No more of the same *pattern* in stock . .	Offlecto
Not much of the same *pattern* in stock .	Offucia
Plenty of the same *pattern* in stock . .	Offula
Price of that *pattern* better quality is — .	Offundo
Price of that *pattern* lower quality is — .	Offusus
Price of that *pattern* same quality is — .	Oggero
Quote price of *pattern* sent in better quality	Oleaceus
Quote price of *pattern* sent in lower quality .	Oleaster
Send at once a selection of *patterns* . .	Oleitas
Pay on our account	Oleosus
Cannot *pay*	Oletum
Cannot *pay* at present	Olfactus
Do not *pay*	Olidus
Do not *pay* at present	Olitor
How do you propose to *pay* . . .	Olivetum
How much is there to *pay*	Olivina
Refuses to *pay*	Olivum
Will *pay* by instalments	Ollaris
Payment must be made against delivery . .	Olorifer
Payment must be made with order . .	Olorinus
Have sold for prompt *payment* . . .	Olus
How is *payment* to be made . . .	Olyra
Permission cannot be obtained . . .	Omasum
Permission has been obtained . . .	Ombria
I give *permission*	Omen
I will not give *permission* . . .	Omentum

Place to my account . . . . . Ominator
 Place to my account at — . . . Ominor
 Place to my account at agent's . . . Ominose
 Place to my account at Bank . . . Omitto
 Place to my account and I will repay you £— Omnifer

Plans have arrived and are approved . . Omnimodo
 Plans have not yet come to hand . . Onager
 Plans submitted will not do . . . Onagrus
 Plans will be sent for approval . . . Onero
 Errors in the *plans* will be corrected . . Onerosus
 Submit *plans* as early as possible . . Onuris

Possession will be given on the — . . . Onustus
 It is not in my *possession* . . . . Opacitas
 It is not in their *possession* . . . Opalia

Post. (Refer to LETTERS, ORDER, WRITING, WROTE.)
 Post is late to-day ; letters not yet delivered Operatio
 Post in a registered letter . . . . Opertus
 Have sent by *post* . . . . . Ophidion
 Have sent by *post* in a registered letter . Ophiusa

Posted letter to-day . . . . . Opicam
 Posted letter to-day, but omitted to en-
  close — . . . . . . Opifer
 Posted letter to-day containing — . . Opimitas
 Posted letter to-day containing cheque ; ac-
  knowledge receipt . . . . Opinator
 Posted letter to-day containing post office
  order; acknowledge receipt . . . Opinor
 Posted letter to-day, do not act before receipt Opipare
 Posted letter to-day, do not act on it . . Opis
 Posted letter to-day, do not act on it, another
  follows . . . . . . Opopanax

Posted letter to-day, do not leave before
receipt . . . . . . Oporice
Posted letter to-day, forward to — . . Oporteo
Posted letter to-day, keep it until — . . Oppango
Posted letter to-day, return it unopened . Oppecto
Posted letter to-day with full instructions . Oppico
Posted letter to-day with full particulars . Opploro
Posted letter to-day with necessary docu-
ments . . . . Oppono
Posted letter yesterday . . . . Oppugno
Posted letter last mail . . . . Optimus
Posting letter by this mail . . . Optio

Post Office order duly received . . . Optivus
Post Office order not yet to hand . . Opulens
Have sent by *Post Office order* . . . Opulus
Send by *Post Office order* . . . . Opuntia
Will send by *Post Office order* . . . Orarius

Poste Restante. (Refer to Letters, Telegram.)
Poste Restante . . . . . . Orarum

Postpone visit, an accident has happened, letter
by post . . . . . . Oratio
Postpone visit for a few days, letter by post. Orbator
Postpone visit on account of illness . . Orbitas
Postpone visit, reasons by letter . . . Orca
Postpone visit until — . . . . Orchis
Postpone visit until to-morrow . . . Orcula
Postpone visit until Monday . . . Ordino
Postpone visit until Tuesday . . . Orexis
Postpone visit until Wednesday . . . Organum
Postpone visit until Thursday . . . Origo
Postpone visit until Friday . . . Orites
Postpone visit until Saturday . . . Ornate
Postpone visit until Sunday . . . Ornatrix

F

Get case *postponed*	Ornatus
Get case *postponed* until arrival of next mail	Ornithon
Get case *postponed,* important evidence by post	Orobitis
Get case *postponed* until —. . . .	Orphus

Power of Attorney must be sent to — . .	Oscito
Power of Attorney sent by post . . .	Osculum
Have given full *Power* to act for me to — .	Ostendo
Have you sent *Power* of Attorney . .	Ostensus
Impossible to act without *Power* of Attorney	Ostentum
You have full *power* to act for me . .	Ostiatim

Premium asked is —	Ostium
What *premium* will be payable . . .	Otiose

Prevent it if possible	Ovatio

Price at present asked is —	Oviaria
Cannot give the *price.*	Ovillus
Send by post *price* of —	Ovis
What is the present *price* of — . . .	Ovum
What *price* will you take	Pabulum

(Private)—This communication is strictly *private*	Pacator

Procure as much as you can	Pacifico
Procure what you want on the spot . .	Paciscor
Cannot be *procured*	Pactilis
Can you *procure* —	Palacra
Will *procure* what you want . . .	Palatum

Profit will be large.	Pallesco
Profit will be small	Palmifer
There will be no *profit*	Palmula
What will be the *profit*	Palpamen

Progress. (Refer to HEALTH.)

Progressing slowly but satisfactorily	. Palpaminum
Making good *progress*	. Palpare
Making little *progress*	. Palpat
Making no *progress*	. Palpamus
What *progress* are you making .	. Palpandum

(Promise)—Can you not *promise* it before —	Palpatio
Can you *promise*	. Palpebra
How soon can you *promise* . .	. Paluster
Unable to *promise*	. Panchrus

Prompt attention is required . .	. Pannosus
Prompt delivery is essential . .	. Panther
Reply as *promptly* as possible . .	. Papilla

Proposal is accepted	. Parabole
Accept the *proposal* . . .	. Paralios
Cannot accept the *proposal* . .	. Parcitas
Do not accept the *proposal* . .	. Parento
Have you any *proposal* to make .	. Parocha
Is the *proposal* accepted. . .	. Paropsis
Proposal entertained, but modifications necessary	. Particeps
Refuses to entertain the *proposal* .	. Parumper
Shall be happy to entertain the *proposal* .	Passim
Shall I accept the *proposal* . .	. Pastinum
Will not entertain the *proposal* .	. Patella
Would a *proposal* from me be entertained .	Patesco

Purchase for me	. Patina
Do not *purchase*	. Patrona
For what can it be *purchased* . .	. Patruus
What amount can you *purchase* .	. Pauci
When will *purchase* be completed .	. Paulatim

F 2

(Purpose)—For what *purpose* do you require . Pavidus

Quality must be guaranteed . . . . Peccatus
   Is the *quality* guaranteed . . . . Pecco
   Quality not equal to sample . . . Pecten
   Must be of the best *quality* . . . Peculio

Quantity on hand is — . . . . . Pedamen
   What *quantity* can you get . . . Pedandus
   What *quantity* do you require . . . Pedes
   What *quantity* have you on hand . . Pedester

(Question)—Cannot answer the *question* . . Pedica
   What is the *question* . . . . . Pegma
   Why have you not answered my *question* . Pelagius

(Quick)—Be as *quick* as possible . . . Pelasga

Racing.
   Acceptances will be published on — . . Pellitus
   — is sure to *accept* . . . . . Peltasta
   — know positively this will not *accept* . Peltiger
   Better be full *against* — . . . . Pelvis
   Impossible to get an offer *against* — . . Penates
   — *Back* this as quickly as possible . . Pendulus
   — *Back* this at starting price . . . Penniger
   — *Back* this for a place . . . . Pensilis
   — *Back* this for a place, best outsider . . Penso
   — *Back* this for double event . . . Pependi
   — is *backed* for genuine money . . . Peplus
   — is being well *backed* by — . . . Peragito
   — is not *backed* for genuine money . . Perasper
   To what extent do you think you can *back* . Perbelle
   — will be well *backed* . . . . Perbibo
   — will probably go *back* in the betting . Perbonus
   Send me latest *betting* . . . . Percaleo

Make your *book* for —	Perceptio
— has *broken down*	Percingo
— has not *broken down* as reported .	Percoquo
— is reported to have *broken down* .	Percrepo
— *cantered* only	Perdisco
— has done a good *canter* to-day .	Perdives
— has been *cast* in his box . .	Perdomo
— reported to have been *cast* in his box .	Perduim
No *change* since last report. . .	Perfecte
Send me latest *changes* . . .	Perferus
Answer when *commission* executed .	Perfidia
Do not execute *commission* in London .	Perflo
Execute *commission* where you please .	Perfrico
— is *coughing*	Perfundo
— is reported to be *coughing* . .	Perfusio
Cover the money laid against — . .	Pergula
is *covering* money only which is going on —	Peritus
Dead-heat. Stakes divided . .	Perlabor
Dead-heat. Will be run off . .	Perlate
— is *disqualified* by death of owner .	Perlonge
— is *disqualified* for — . . .	Perlubet
Winner was *disqualified,* could not draw weight.	Permano
Winner was *disqualified* for carrying over-weight.	Permotus
Winner was *disqualified* for crossing .	Permunio
Winner was *disqualified* for foul riding .	Pernix
Do nothing until you hear from me again .	Pernocto
— is a *doubtful* runner . . .	Perpaco
Do not *fancy* —	Perparum
There will be a large *field* . . .	Perpendo
There will be a small *field* . . .	Perplexe
It is reported the *field* will be large .	Perpolio
It is reported the *field* will be small .	Perprimo
— has done a good *gallop* . . .	Perrogo
Answer if you cannot *get on* . .	Persea

— has *gone* to —	Persisto
Am told this is a *good* thing for — .	Perspexi
— is *good*	Persuasi
Hedge quickly all you have against —.	Persulto
Hedge quickly all you have on — .	Pertendo
— has cracked *heel*	Pertexo
— has been sent *home* . . .	Pertinax
— will be *knocked* out . . .	Pervinco
Do you *know* anything for — . .	Pervius
— dead-*lame* after exercise . .	Pestifer
— is *lame*	Petalium
— is reported *lame*	Petaso
Lay the odds to £5 against — .	Petiolus
Lay the odds to £10 against — .	Pexatus
Lay the odds to £15 against — .	Phani
Lay the odds to £20 against — .	Phaselus
Lay the odds to £25 against — .	Phasma
Lay the odds to £30 against — .	Phellos
Lay the odds to £35 against — .	Phiditia
Lay the odds to £40 against — .	Phormion
Lay the odds to £45 against — .	Phragmis
Lay the odds to £50 against — .	Phycos
Lay the odds to £55 against — .	Phylaca
Lay the odds to £60 against — .	Phyllon
Lay the odds to £70 against — .	Physema
Lay the odds to £80 against — .	Piaculum
Lay the odds to £90 against — .	Pignero
Lay the odds to £100 against — .	Pigritia
Lay the odds to £150 against — .	Pinaster
Lay the odds to £200 against — .	Pineta
Lay the odds to £250 against — .	Pistacia
Lay the odds to £300 against — .	Pityusa
Lay the odds to £400 against — .	Placenta
Lay the odds to £500 against — .	Placidus
Lay the odds to £600 against — .	Plagiger
Lay the odds to £700 against — .	Plagosus

Lay the odds to £800 against — . .	Planipes
Lay the odds to £900 against — . .	Plantago
Lay the odds to £1,000 against — . .	Pleiades
Lay the odds to £2,000 against — . .	Plinthus
Lay the odds to £3,000 against — . .	Plorator
Lay the odds to £4,000 against — . .	Plumatus
Lay the odds to £5,000 against — . .	Plumbo
Lay over your book against — . . .	Plumesco
Lay your book only against — . . .	Pluvius
Cannot *lay* at any price	Podager
Cease *laying* against —	Podex
Good men are *laying* —	Podium
Market arranged, do not *lay* . . .	Poetica
Succeeded in *laying* for you — . . .	Pogonias
— has hit its *leg*	Politicus
— reported to have *leg* filled . . .	Polygala
— is only a *market* horse	Pomosus
— has the *mount*, and fancies it . .	Pompa
— has the *mount*, but does not fancy it .	Pondero
Money is right with —	Ponto
Money is wrong with —	Popina
Answer what *odds* you have obtained . .	Poples
— has not been *out* to-day . . .	Populus
— is a good *outsider*	Portatus
— is the best *outsider*	Positio
Owner cannot get on, public got the money .	Possedi
Owner does not fancy	Possunt
Owner fancies —	Postmodo
— is sure to get a *place*	Postpono
Put me £1 to win on —	Postquam
Put me £1 to win, and £1 for a place on —	Potior
Put me £2 to win on —	Potorius
Put me £2 to win, and £2 for a place on —	Pransito
Put me £3 to win on —	Pratulum
Put me £3 to win, and £3 for a place on —	Pravitas
Put me £4 to win on —	Precario

Put me £4 to win, and £4 for a place on — .	Prehendo
Put me £5 to win on —	Pretiose
Put me £5 to win, and £5 for a place on —	Pridianus
Put me £6 to win on —.	Primum
Put me £6 to win, and £6 for a place on —	Princeps
Put me £7 to win on —	Privatim
Put me £7 to win, and £7 for a place on —	Proavia
Put me £8 to win on — . ' . . .	Probatio
Put me £8 to win, and £8 for a place on —	Problema
Put me £9 to win on —	Probum
Put me £9 to win, and £9 for a place on —	Procax
Put me £10 to win on —	Proclamo
Put me £10 to win, and £10 for a place on—	Proculco
Put me £20 to win on —	Procumbo
Put me £25 to win on —	Procuro
Put me £30 to win on —	Prodige
Put me £35 to win on —	Prodere
Put me £40 to win on —	Proditur
Put me £45 to win on —	Profano
Put me £50 to win on —	Profaris
Put me £60 to win on —	Profindo
Put me £70 to win on —	Prohibeo
Put me £75 to win on —	Prolabor
Put me £100 to win on —.	Proles
Put me £200 to win on —. . . .	Prolixe
Put me £300 to win on —. . . .	Prolixus
Put me £400 to win on —. . . .	Proludo
Put me £500 to win on —. . . .	Prompto
Put me £1,000 to win on — . . .	Promulgo
What do you *recommend*	Pronecto
Am waiting here for your *reply* . . .	Pronepos
— has been *retained* to ride — . . .	Pronomen
— is all *right*, no cause for apprehension .	Pronubus
— *runs*, and is meant to win	Propalam
— is sure not to *run*	Propello
— is sure to *run*	Propense

— will *run*, but do not fancy it . . .	Propior
The number of *runners* will be at least — .	Propola
The number of *runners* will probably be about —	Proposui
— should be kept on the *safe side* . .	Proprie
— is *scratched*	Proptosis
— is not *scratched* as reported . . .	Propulso
— is reported *scratched*	Proreta
Succeeded in *taking* for you — . . .	Prorogo
Trial was genuine	Proserpo
Trial was not genuine	Prosilio
— has been *tried*, and beaten . . .	Prospere
— has been *tried*, and beaten by — . .	Prospexi
— has been *tried*, and won easily, beating —	Prosterno
— has been *tried*, and won, stable is satisfied	Prosto
— took *walking* exercise only . . .	Prosum
Think something is *wrong* with — . .	Protelum

Races, List of Important.

N.B.—All Races having names limited to one word, such as Derby, Oaks, Cesarewitch, &c., are purposely omitted in this list.

Alexandra Plate, *Ascot*	Protendo
Ascot Gold Cup	Proterve
Champagne Stakes, *Doncaster* . . .	Protollo
Chesterfield Cup, *Goodwood* . . .	Protraho
Chesterfield Stakes, *Newmarket* . . .	Protrudo
City and Suburban Handicap . . .	Protypum
Earl Spencer's Plate, *Northampton* . .	Provenio
French Derby	Providus
French Oaks	Provoco
Gold Cup, *Epsom*	Proxime
Goodwood Cup	Pruina
Goodwood Stakes	Prunum
Grand Prix de Paris	Pruritus

Great Challenge Stakes (*Newmarket* Second October)	Prytanis
Jockey Club Cup (*Newmarket Houghton*) .	Psaltria
Liverpool Grand National	Psegma
Middle Park Plate	Psora
New Stakes, *Ascot*	Psyllion
Newmarket October Handicap . . .	Pucinum
One Thousand Guineas	Pudicum
Prince of Wales's Stakes, *Ascot* . . .	Puellas
Rous Memorial Stakes, *Ascot* . . .	Pueritia
Royal Hunt Cup, *Ascot*	Pugillar
St. Leger Stakes, *Doncaster* . . .	Pugillum
Two Thousand Guineas	Pugnator

Railway. (Refer to ACCIDENT, ENQUIRE, PARCEL, SEND, TRAIN.)

Ready at a moment's notice	Pulcher
Am quite *ready* to start	Pulecium
Please hold yourself *ready* to start . .	Pulex
When will you be *ready*	Pullulo

Reduction is asked to the extent of — . .	Pulmo
Endeavour to get a *reduction* . . .	Pulsatio
What *reduction* can be made . . .	Pulvinar
What *reduction* is asked for . . .	Pumilus

(Refer)—You may *refer* to	Punctus

References are satisfactory	Punitio
References are not satisfactory . . .	Purifico
Our *references* are —	Pustula
Further *references* are necessary . . .	Putamen
What *references* can you give . . .	Putealis

Remain at home	Putredo

Remain at home for me	Putresco
Remain at home, will call this evening .	Putror
Remain there for the present . . .	Pyralis
Remain there until I come. . . .	Pyrgus
Remain there until —	Pyrites
Am I to *remain* here	Pyropus
Better *remain* —	Pythia
Better *remain* a few days	Pythicus
Better *remain* another day	Rabidus
Better *remain*, as you suggest . . .	Rabiose
Better *remain* there for the present . .	Radicor
How long am I to *remain*	Radicula
How long will you *remain*	Radiosus
Likely to *remain* here a few days . .	Radius
Likely to *remain* here another week . .	Ramale
Likely to *remain* here another fortnight .	Ramex
Likely to *remain* here another month . .	Ramosior
Likely to *remain* here to-day . . .	Ranceo
Likely to *remain* here until — . . .	Raptio
Must *remain* here until —	Raritas

Remittance. (Refer to Acknowledge, Cheque, Money, Place.)

Remittance has been sent as requested . .	Raritatis
Remittance is not yet to hand . . .	Raritudo
Remittance is to hand.	Rasito
Cannot *remit* before —	Rastrum
Cannot *remit* more than —	Raucitas
Have *remitted* as requested £ — . .	Ravem
Have *remitted* to the credit of — . .	Ravis
What amount have you *remitted*. . .	Reapse
When will you *remit*	Reatus

Renew. (See Acceptance.)

Repairs are in progress, but not yet completed . Rebellis

Repairs will be completed by — . .	Recalco
Can it be *repaired*	Recalvus
Can you *repair*	Recenter
Damage is *repaired*	Receptio
Do not do any *repairs*	Receptus
Shall I have it *repaired*	Rechamus
What *repairs* are required	Recingo
What will the *repairs* cost	Recinium
When will *repairs* be completed . . .	Recito

Reply. (Refer to LETTERS, ORDER, TELEGRAM.)

Reply as soon as possible	Reclivis
Reply by letter	Recludit
Reply by messenger	Recludo
Reply by telegram	Recludunt
Reply cannot be given for a few days . .	Reclusi
Reply shall be sent by post to-day . .	Reclusus
Reply shall be sent by telegram . . .	Recogito
Reply shall be sent in a few days . .	Recoquo
Reply shall be sent next mail . . .	Recordor
Reply shall be sent to-morrow . .	Recreatum
Cannot understand why you do not *reply* .	Recreavi
Reply has not yet come to hand . . .	Recupero

Report is not correct . . . . . Recurso
   Report is quite correct . . . . Recurvus
   Send further *report* immediately . . . Redactus

(Represent)—Will *represent* me in the matter . Redamo

Request cannot be complied with . . . Redarguo
   Request will be complied with . . . Redditio

Resolution was carried by a large majority . Redhibeo
   Resolution was carried by a small majority . Redimio

Resolution was carried unanimously . . **Reditum**
Resolution was lost by a large majority . **Redoleo**
Resolution was lost by a small majority . **Redormio**
It has been *resolved* that — . . . **Redundo**

(Responsibility)—Accept the *responsibility* . **Reduvia**
Cannot accept the *responsibility* . . . **Redux**
Refuses to be *responsible* . . . . **Refectio**
Who is *responsible* . . . . . **Refector**
Who will be *responsible* . . . . **Refercio**
You are *responsible* . . . . . **Reflatus**
You are not *responsible* . . . . **Reflecto**

Result is in favour of — . . . . **Refodio**
Result is not yet known . . . . **Reformo**
Result is satisfactory . . . . . **Refragor**
Result is unsatisfactory . . . . **Refregi**
Result will not be known for a few days . **Refringo**
Result will probably be in favour of — . **Refugium**
Let me know the *result* of — . . . **Refugus**
Let me know the *result* of the election . **Regalis**
Let me know the *result* of the match . . **Regelo**
Let me know the *result* of the meeting . **Regie**
Let me know the *result* of the race . . **Regifice**
Let me know the *result* of your enquiries . **Regimen**
Let me know the *result* of your interview . **Regno**

Return at once . . . . . . **Regnum**
Return at once, all arranged satisfactorily . **Regulus**
Return at once, all shall be arranged as you
desire . . . . **Regusto**
Return at once, or consequences will be
serious . . . . . . **Regyro**
Better *return* — . . . . . **Rehalo**
Better *return* to-day . . . . . **Reice**
Better *return* to-morrow . . . . **Rejectus**

Shall *return* in time for dinner . . . Relabor
Shall *return* in time for dinner, am bringing — Relatio

Sample has arrived. . . . . . Relaxus
Sample has been sent. . . . . Relictum
Sample has not arrived . . . . Relino
Sample will be sent . . . . . Reliquus
Forward another *sample* . . . . Reluctor
Forward by post *sample* and price of — . Remano
When will *sample* be sent . . . . Remansio

Security offered is accepted . . . . Remex
Security offered is not accepted . . . Remigium
' Better *security* is required . . . . Remisceo
Can you obtain good *security* . . . Remissus
Get the best *security* you can . . . Remolior
What *security* have you obtained . . Remoram

(Sell)—At what price may I *sell* . . . Remordeo
Do not *sell* at any price . . . . Remotus
Do not *sell* at less than — . . . Remugio
Do not *sell* until further orders . . . Remunero

Send as early as possible . . . . . Renarro
Send at once by — . . . . . Renatus
Send at once by cheque . . . . Renavigo
Send at once by quickest means — . . Renendus
Send at once by P.O.O. — . . . Renixus
Send at once by Parcel Post . . . Reno
Send at once by post . . . . . Renuntio
Send at once by railway goods train . . Renuntius
Send at once by railway passenger train . Repandus
Send at once by Great Eastern Railway Co. Reparco
Send at once by Great Northern Railway Co. Repecto
Send at once by Great Western Railway Co. Repente

Send at once by London, Brighton, and South Coast Railway Co. . . .	Repertor
Send at once by London, Chatham, and Dover Railway Co.	Repexus
Send at once by London and North-Western Railway Co.	Repletus
Send at once by London and South-Western Railway Co.	Replico
Send at once by Midland Railway Co. .	Replum
Send at once by South Eastern Railway Co.	Replumbo
Send at once by special messenger . .	Repono
Send at once with the goods already on order	Reportare
Send authority to —	Reporto
Send authority for —	Repostus
Send authority at once	Repotia
Send for enclosure to-day to — . . .	Repotiorum
Send for enclosure to-morrow to — . .	Repressi
Send full instructions.	Repressor
Send full particulars of —	Reprobo
Send full particulars of accident . . .	Reptatio
Send full particulars of claim . . .	Repudio
Send full particulars of damage . . .	Repudium
Send my evening dress here . . .	Repugno
Send my evening dress to — . . .	Repulsus
Send my flannels here	Repurgo
Send my flannels to —	Reputo
Send to our order at Railway Station at —.	Reputabo
Send to our order at Railway Station here .	Reputans
Send to our order at wharf here . . .	Reputavi
Have *sent* as requested	Requies
Have *sent* by special messenger . . .	Requiro
Have *sent* by Great Eastern Railway Co. .	Resaluto
Have *sent* by Great Northern Railway Co. .	Resarcio
Have *sent* by Great Western Railway Co. .	Resecro

Have *sent* by London, Brighton, and South
Coast Railway Co. . . . . Resectus
Have *sent* by London, Chatham, and Dover
Railway Co. . . . . . Resedo
Have *sent* by London and North-Western
Railway Co. . . . . . Resegmen
Have *sent* by London and South-Western
Railway Co. . . . . . Resemino
Have *sent* by Midland Railway Co. . . Resequor
Have *sent* by South-Eastern Railway Co. . Reservo

(Service) — Will you take *service* for me
on — . . . . Residuus
Will you take *service* for me on Sunday ' . Resignat
Will you take *service* for me to-morrow . Resilio

(Settle)—Cannot *settle* at present, will do so
shortly . . . . . . Resistit
Everything *settled*, return at once . . Resolvo
Everything *settled*, telegraph when you will
come . . . . Resonus
Settlement arrived at satisfactorily . . Resorbeo
Settlement is impossible . . . . Resorpsi
Settlement must be come to . . . Respergo
Settlement must be come to immediately . Respiro
Settlement must be come to by — . . Restagno

Ship is detained in quarantine . . . Restillo
Ship is detained in port . . . . Restiti
Ship is due on the — . . . . Restringo
Ship is just leaving port . . . . Resudo
Ship sailed from here on the — . . . Resulto
Ship will sail on the — . . . . Resupino
What is the name of the *ship* . . . Resurgo
When did the *ship* arrive . . . . Retardo
When was the *ship* last heard of . . . Retectus

(**Size**)—What is about the *size* . . . Retego
What *size* do you want . . . . Retentio

Sold by order of the Court . . . . Retexui
Sold by public auction . . . . Reticeo
Sold with all faults . . . . . Retono
For how much has it been *sold* . . . Retostus
It has been *sold* for — . . . . Retribuo
Was not *sold* . . . . . . Retroago
Will not be *sold* . . . . . Retrudo

(**Standstill**)—Am quite at a *standstill* . . Retrusus

(**Start**)—Cannot *start* at time agreed on . . Reunctor
Cannot *start* to-day . . . . . Reus
Cannot *start* until —. . . . . Reveho

Statement is confirmed . . . . . Revera
Statement is denied . . . . . Reversus
Statement is incorrect, send another . . Revincio
Full *statement* sent by post . . . Revisito
Send *statement* of account . . . . Revocatio
You are authorised to deny the *statement* . Revocare

Stock is abundant . . . . . . Revulsi
Stock is very low . . . . . Rhacinus
What have you in *stock* . . . . Rhacoma

Subject to a discount of — . . . . Rhagion
Subject to analysis . . . . . Rhamnus
Subject to your approval . . . . Rhetor

Sufficient time must be given . . . . Rhexia
Is not *sufficient* . . . . . Rhodora
Is quite *sufficient* . . . . . Rhoicus

G

(Suit)—It will not *suit* . . . . . . Rhombus
    It will *suit* very well . . . . . . Rhyas

Supply is exhausted . . . . . Rhytion
    Send a further *supply* of — . . . Rigatio

Sympathise deeply with you in the loss you
    have sustained . . . . . Rigesco
    Sympathise deeply with you in your trouble . Rigide

(Take)—Do not *take* — . . . . . Rigidor
    How long will it *take* . . . . Rigoris
    How long will it *take* to complete . . Riparius
    Refuse to *take* it back . . . . Ritualis
    When will it *take* place . . . . Rixator
    Will *take* it into consideration . . . Rixosus
    Will *take* it with me . . . . . Roboreus

Telegram. (Refer to ACKNOWLEDGE, ADDRESS,
    CANCEL, HOTELS, ORDER, UNICODE,
    WAIT.)
    Telegram received . . . . . Roboro
    Telegram received, agree to — . . . Robur
    Telegram received, agree to your terms . Robustus
    Telegram received, cannot agree to contents . Rodo
    Telegram received, cannot agree to terms,
    will write . . . . . . Rogalem
    Telegram received, cannot agree to your
    terms . . . . Rogator
    Telegram received, cannot cancel orders . Rogito
    Telegram received, cannot cancel orders,
    already attended to . . . . Roresco
    Telegram received, cannot do so . . . Rorifer
    Telegram received, cannot do so to-day. . Rosetum
    Telegram received, cannot meet you . . Rosmaris

Telegram received, cannot meet you until —	**Rotatus**
Telegram received, cannot understand its meaning, wire again in different words .	**Rotula**
Telegram received, have done as you requested	**Rotundus**
Telegram received, have written you fully .	**Rubellus**
Telegram received, meeting postponed . .	**Rubesco**
Telegram received, meeting postponed until —	**Rubetum**
Telegram received, orders cancelled . .	**Rubia**
Telegram received, orders cancelled, and subsequent ones substituted . . .	**Rubrica**
Telegram received too late	**Rudis**
Telegram received too late for — . .	**Rufulus**
Telegram received too late for post . .	**Rugosus**
Telegram received, will meet you . .	**Ruidus**
Telegram received, will meet you at — .	**Ruiturus**
Telegram received, will meet you as desired .	**Rumpus**
Telegram received, will do as you wish .	**Runcator**
Telegram and letter received . . .	**Runcina**
Telegram and letter received, having attention	**Runco**
Telegram and letter received, having attention, will write	**Ruptor**
Our last *telegram* was dated — . . .	**Ruralis**
Reply immediately to *telegram* sent on — .	**Rursum**
There is no *telegram* from you here, wire to me at once to —	**Rurum**
There is no *telegram* from you here, wire to me at once to this address . .	**Ruscario**
There is no *telegram* from you here, wire to me to Poste Restante at — . .	**Ruscarium**
There is no *telegram* from you here, wire to me to Poste Restante here . .	**Ruscum**
Telegraph date despatched	**Russatus**
Telegraph date of departure . . .	**Rustice**
Telegraph him [or her] at Poste Restante — .	**Rusticitas**
Telegraph how many are wanted . .	**Rusticor**
Telegraph me at Poste Restante — . .	**Ruta**

G 2

Telegraph me at Poste Restante here . .	**Rutarum**
Telegraph present price of — . . .	**Sabulo**
Telegraph reply to letter	**Sacal**
Telegraph reply to letter of — . . .	**Sacco**
Telegraph result of —	**Sacculus**
Telegraph result of your interview . .	**Sacellum**
Telegraph substance of your letter . .	**Sacodios**
Telegraph what progress you are making .	**Sacoma**
Telegraph your address for letters to be posted to you to-night	**Sacomatis**
Telegraph your address for letters to be posted to you to-morrow . . .	**Sacrarium**
Telegraph your address for letters to be posted to you on —	**Sacratum**

Terms. (Refer to Telegram.)

Terms are accepted	**Sacrifer**
Terms are considered satisfactory. . .	**Sacris**
Terms are rejected . . . .' .	**Sagatus**
Terms are too high	**Sagdarum**
Terms are too low	**Saginatio**
Terms will not suit	**Sagminis**
Accept the *terms* offered	**Sagulum**
Cannot accept other *terms* than already named	**Sagum**
Get the best *terms* possible	**Salacia**
Get the best *terms* possible, and wire result .	**Salarius**
What are the best *terms*	**Salebra**
What are your *terms*	**Salgama**
What *terms* are agreed to	**Salictum**

(Thanks)—Accept our best *thanks* . . . **Salignus**

Theatre, or Concert.

Book a private box for the — . . .	**Salillum**
Book a private box for this evening . .	**Salina**
Book one stall for —	**Saliunca**

Book two stalls for —	Sálivosus
Book three stalls for —	Salsius
Book four stalls for —	Salsura
Book one dress circle seat for — . .	Saltatio
Book two dress circle seats for — . .	Saltem
Book three dress circle seats for — . .	Saluber
Book four dress circle seats for — . .	Saluto
Book one upper circle seat for — . .	Sambuca
Book two upper circle seats for — . .	Samiolus
Book three upper circle seats for — . .	Samnites
Book four upper circle seats for — . .	Sandalis
Cannot get tickets	Sandyx
Cannot get tickets you want . . .	Sanesco
Cannot get tickets you want, shall I take for —	Sangenon
Have a private box for to-night for the — .	Sanguino
Have stalls for to-night for the — . .	Sanitas
Have dress circle tickets for to-night for the	Santerna
Have upper circle tickets for to-night for the	Saperda
Have tickets for to-night for the — . .	Sapineus
Received *invitation* for concert this evening at—	Sapros
Received *invitation* for concert this evening, can you come	Sarcina
Received *invitation* for theatre this evening .	Sardonyx
Received *invitation* for theatre this evening, can you come	Sarissa
Received *invitation* for theatre this evening, meet me at —	Sarmen
Shall I take a box for the — . . .	Sarritio
Shall I take stalls for the — . . .	Sartago
Shall I take tickets for the — . . .	Satageus
Take a box for to-night for the — . .	Satagito
Take stalls for to-night for the — . .	Satisdo
Take dress circle seats for to-night for the —	Sativus
Take upper circle seats for to-night for the —	Satura
Take tickets for to-night for the — . .	Saucius

There ought to be	Saxifer
There ought not to be	Scaber
There will be	Scabres
There will not be	Scala
Time has already expired	Scalprum
Time is too long	Scambus
Time is too short	Scamnum
Time must be made the essence of the contract	Scando
Time will expire on the —	Scansilis
Can you alter the *time* to — . . .	Scarifico
Can you extend the *time* for — . . .	Scarites
Shall I be in *time* for —	Scatebra
There is not sufficient *time*	Scena
There will be plenty of *time* . . .	Scheda
What is the latest *time* for — . . .	Schidius
Too long to telegraph details, am writing fully .	Schiston
(Train)—Have missed *train*	Scholium
Have missed *train,* cannot arrive in time for —	Scienter
Have missed *train,* cannot arrive this evening	Scilicet
Have missed *train,* cannot arrive until — .	Scincus
Have missed *train,* do not expect me . .	Scio
Have missed *train,* impossible to be home before —	Scirpeus
Have missed *train,* impossible to be with you to-night	Scirroma
Have missed *train,* make other arrangements	Scloppus
Have missed *train,* postpone meeting . .	Scobina
Have missed *train,* postpone meeting until—	Scolymos
Have missed *train,* remaining here to-night	Scombrus
Have missed *train,* send carriage to meet me at —	Screatus

Have missed *train*, send conveyance to meet me at —	Scrinium
Have missed *train*, shall arrive later . .	Scriptio
Have missed *train*, unable to keep appointment	Scrofula
Have missed *train*, wait for me until — .	Scutatus
Have missed *train*, wait until I arrive .	Scutella
Have missed *train*, will come by first in the morning	Scutica
Leaving by *train* arriving at — . . .	Scyphus
Leaving by *train* arriving at Cannon Streèt Station at —	Scyricum
Leaving by *train* arriving at Charing Cross Station at —	Ṣcytala
Leaving by *train* arriving at Euston at — .	Scythica
Leaving by *train* arriving at Fenchurch Street at —	Secedo
Leaving by *train* arriving at Holborn Viaduct at —	Secessio
Leaving by *train* arriving at King's Cross at —	Secius
Leaving by *train* arriving at Liverpool Street at —	Secludo
Leaving by *train* arriving at London Bridge at —	Secretio
Leaving by *train* arriving at Paddington at —	Secundum
Leaving by *train* arriving at St. Pancras at —	Sedecula
Leaving by *train* arriving at Victoria at —.	Seduco
Leaving by *train* arriving at Waterloo at —	Segestre
Leaving by *train* due at — . . .	Segnipes
Leaving by *train*, meet me at —. . .	Segnis
Leaving by *train*, meet me at station at — .	Segnitia
Leaving by *train*, send carriage to meet me at —	Segrego

Leaving by *train*, send conveyance to meet
me at — . . . . . . Segullum
Leaving by *train*, shall be with you at — . Sejungo
Leaving by *train* this afternoon . . . Selago
Leaving by *train* this evening . . . Selectio
Leaving by *train* this morning . . . Selibra

(**Trial**)—When will the *trial* take place . . Sella

(**Trouble**)—Do not *trouble* in the matter . . Sellaria
Will necessitate too much *trouble* . . Sementis

(**True**)—It is not *true* . . . . . Semibos
It is quite *true* . . . . . . Semidea

(**Trusted**)—Are they to be *trusted* . . . Semihora
Can be *trusted* to the extent of — . . Seminex
Do not *trust* — . . . . . Semito
To what extent can they be *trusted* . . Semodius

(**Understand**)—Do not *understand* your letter . Semoveo
Do not *understand* your telegram . . Semuncia
Do you *understand* . . . . . Semustus
Do you *understand* our meaning . . . Senecio
Does he *understand* . . . . . Senectus

(**Unicode**)—To decipher this message refer to
the Unicode . . . . . Unicode

(**Unnecessary**)—Consider it *unnecessary* . . Senex

(**Unsaleable**)—Is *unsaleable*, unless at a heavy loss Senium
Is quite *unsaleable* . . . . . Sentisco

Unsatisfactory reports have arrived . . Seorsum
Is very *unsatisfactory* . . . . Sepelio

Visit.   (Refer to Appointments, Postpone, Train.)

Wait for me at — . . . . . . Septicus
   Wait for me this evening . . . . Sereno
   Wait for me this evening, will call . . Sergia
   Wait my arrival . . . . . Seriola
   Wait my letter before starting . . . Serius
   Wait my letter before taking any action . Serratim
   Wait my telegram before starting . . Sesqui
   Wait my telegram before taking any action Sessito
   Wait until you hear further before —. . Sestiana
   Wait until you receive my letter . . Setanium
   Am *waiting* here for a letter from you before
     starting . . . . . . Setosus
   Am *waiting* here for a telegram from you
     before starting . . . . . Sevoco

Weather too unfavourable . . . . Sexatruus
   Weather too unfavourable, do not come . Sextiana
   Weather too unfavourable, must postpone — Sextula
   Weather too unfavourable, return by rail . Sexus
   Weather too unfavourable, returning by next
     train . . . . Siccanus
   Weather too unfavourable to put to sea . Siccine
   Weather too unfavourable to put to sea, will
     telegraph departure . . . . Siderior
   Weather too unfavourable to start to-day . Sido
   Weather very fine . . . . . Sigma
   Weather very fine, excursion to-day . . Signifer
   Weather very fine, sea quite smooth . . Signinus
   Weather very fine, shall expect you—. . Signum
   Weather very fine, shall expect you this
     morning . . . . Silaceus
   Weather very fine, shall expect you this
     afternoon . . . . . . Siliqua

Weather very fine, shall expect you this evening	Silurus
Weather very fine, shall start to-day . .	Silva
Weather very fine, will wait your arrival .	Silvesco

When did you last hear from — . . . Silvicola

Will. (Refer to DEATHS, EXECUTORS.)

Writing. (Refer to LETTER, POST.)

Writing to you by to-day's post . . .	Similago
Writing to you by early post . . .	Similis
Writing to you by next mail — . . .	Simplex
Writing you to-day respecting — . .	Simultas
Writing you to-morrow respecting — . .	Sinciput

(Wrong)—Is anything *wrong* . . . . Sindon

Is anything *wrong*, have received nothing
from — . . . . . . Singulus

Is anything *wrong*, have not heard from you Sinister

Is anything *wrong*, have not heard from you
for some time . . . . . Sinopis

Nothing *wrong*, will write . . . . Sinum

Wrote. (Refer to LETTER, POST.)

Wrote to you addressed to — . . .	Sinuosus
Wrote to you by mail of last — . . .	Siparium
Wrote to you by this evening's post . .	Siquandare
Wrote to you by this morning's post . .	Siquando

PRIVATE CODE.

A SIMPLE means of converting the Unicode into a secret private code is for correspondents to arrange to use instead of the cypher set opposite to the phrases in the book the cypher affixed to the phrase one, two, or more lines above or below, as may be selected. For instance, if it is agreed to use instead of the regulation cypher word the one next following it in the Code, a telegram with the word "*Oporice*" would mean to the general body of Unicode users "Posted letter to-day, do not leave before receipt;" but the person for whose private information the message was intended would read the real meaning as "Posted letter to-day, do not act on it, another follows."

The following selection of cypher words will never be included in the "Unicode" for permanent use with any specific phrases. They are intended to be used only for private phrases to be arranged specially between individual correspondents :—

Veneno

Venenum

Venereus

Veneror

Venetus

Venicula

Venor

Venosus

Ventilo

Ventrale

" *UNICODE*":

Venundo

Venus

Vepres

Vepretum

Veratrum

Verax

Verbena

Verber

Verbosus

Veredus

Veretrum

Veritas

Vermino

Vermis

Vernatio

Verpus

Verrinus

Verruca

Versoria

Vertagus

Vertebra

Vertigo

Verum

Verutum

Vesania

Vescor

Vesica

Vesicula

Vesper

Vestio

Vestras

Veto

Vetulus

Vetustas

Vexatio

Vexator

Vexillum

Vialis

Viarius

Viaticus

Viator

Vibex

Vibro

Vicarius

Vicatim

Vicinia

Vicissim

Victima

Victito

Victrix

Vidi

Viduitas

Vidulum

Viduus

Vigesco

Vigil

Vigilax

Vigilia

Vigor

Villa

Villaris

Villicus

Villula

Vimen

Vinaceus

Vinca

Vincio

Vinctura

Vincales

Vindemia

Vindex

Vindico

Vinetum

Vinitor

Viola

Violator

Violens

Violenter

Vipera

Vipereus

Virago

Viresco

Viretum

Virgula

Viridis

Viritim

Virosus

Virtus

Visula

Vix

The following cypher words have been added to " Unicode " since its first publication :—

Antidotum	Obvigilo
Anxiferum	Obviam
Congruum	Obvolvunt
Conifer	Occallatus
Coniferum	Occano
Mellilla	Occasiones
Mellinia	Orarum
Mellis	Palpaminum
Mellitorum	Palpare
Mellitum	Palpat
Mello	Palpamus
Mellorum	Palpandum
Mellum	Raritatis
Melofolia	Ravem
Melofoliam	Recludit
Meminisse	Recludo
Memoris	Recludunt
Memoratus	Reportare
Obvertunto	Repotiorum
Obvigilat	Repressi
Obvigilavi	Reputabo

H

' *UNICODE.* "

Reputans	Rutarum
Reputavi	Sacomatis
Rotula	Sacrarium
Rurum	Sacratum
Ruscario	Sinuosus
Ruscarium	Siparium
Ruscum	Siquandare
Rusticitas	Siquando
Ruta	

UNICODE USERS.

The following is a List of important firms and establishments to whom messages in the "Unicode" may be sent by any persons at any time without necessity for previous arrangement. Their registered telegraphic address is also given.

NAME OF FIRM.	REGISTERED TELEGRAPHIC ADDRESS.
Addams-Williams, R., 16, Commercial Street, Newport, Monmouth (and at Crickhowell)	Addams-Williams, Newport, Mon.
Alabaster, Passmore & Sons, Fann Street, Aldersgate Street, London	Alamores, London.
Allan Brothers & Co., Allan Royal Mail Line, 103, Leadenhall Street, London	Allanline, London.
"Anchor" Line (see Henderson Brothers).	
Anglo-American Brush Electric Light Corporation, Limited, Belvedere Road, London	Magneto, London.
Anglo-American Rope and Oakum Company, 12, Hopwood Street, Liverpool	Oakum, Liverpool.
Army and Navy Co-operative Society, Limited, 117, Victoria Street, London	Army, London.
Arnold, E. J., 3, Briggate, Leeds	Arnold, Leeds.
Artistic Stationery Co., Limited, Plough Court, Fetter Lane, London	Artistic, London.
Bain, W., & Co., 6a, Victoria Street, Westminster, London. Edinburgh	Lochrin, London. Lochrin, Edinburgh.
Baird, William & Co., 168, West George Street, Glasgow	Bairds, Glasgow.
Ballantyne, Hanson & Co., 4, Chandos Street, Charing Cross, London	Ballantyne Press, London.
Edinburgh : Paul's Work	Ballantyne Press, Edinburgh.
Bemrose & Sons, 23, Old Bailey, London. Derby	Bemrose, London. Bemrose, Derby.
Binnie, James, 69, Bath Street, Glasgow	Gartcosh, Glasgow.
Binns, Walter, Stanhope Works, Horton Lane, Bradford	Stanhope, Bradford.
Bird, Bookseller, Tring	Bird, Tring.
Bollans, E. & Co., Ranelagh Works, Leamington	Bollans, Leamington.
Brandauer, C., & Co., Steel Pen Works, New John Street West, Birmingham	Brandauer, Birmingham.
Brendon, William, & Son, Printers, Plymouth	Brendonson, Plymouth.
Brinsmead, John, & Sons, 18, 20, and 22, Wigmore Street, London	Brinsmead, London.
Brown's Hotel (J. J. Ford & Sons), 22, Dover Street, London	Brownotel, London.
Brown, Scott & Co., Red Lion Yard, 254, High Holborn, London	Punctual, London.
Brown, A., & Sons, 26 and 27, Savile Street, Hull	Brown, Hull
Bull, William James, 21, Westcroft Square, Ravenscourt Park, Hammersmith, London; National Union Club, Albemarle Street ; and London Sailing Club.	

NAME OF FIRM.	REGISTERED TELEGRAPHIC ADDRESS.
Bullivant & Co., 72, Mark Lane, London	Bullivants, London.
Caldwell Brothers, Limited, Waterloo Stationery and Printing Works, 11, 13, and 15, Waterloo Place, Edinburgh	Caldwells, Edinburgh.
Callender's Bitumen, Telegraph and Waterproof Co., Limited, 101, Leadenhall Street, London . .	Callender, London.
Cameron & Ferguson, 88, West Nile Street, Glasgow .	Exemplum, Glasgow.
London : Salisbury Court, Fleet Street . . .	Cameronius, London.
Cammell, Charles, & Co., Limited, Cyclops Steel and Iron Works, Sheffield	Cammell, Sheffield.
Campbell, P. and P., The Perth Dye Works, Perth .	Campbell, Perth.
Carter, F. & F. W., Chartered Accountants, 5, St. Andrew's Square, Edinburgh	Carter, Edinburgh.
Cassell & Co., Limited, La Belle Sauvage, Ludgate Hill, London	Caspeg, London.
Causton, Sir Joseph, & Sons, 9, Eastcheap, London .	Caustcheap, London.
Chatwood's Patent Safe and Lock Co., Limited, 76, Newgate Street, London	Chatwoods, London.
Manchester : 11, Cross Street	Chatwoods, Manchester.
Liverpool : 17, Lord Street	Chatwoods, Liverpool.
Leeds : 22, Bond Street	Chatwoods, Leeds.
Bolton : Lancashire Safe and Lock Works . .	Chatwoods, Bolton.
Chubb & Son's Lock and Safe Company, Limited, 128, Queen Victoria Street, London. . .	Chubb, London.
City Timber & Saw Mills Co., Limited, 168, London Road, Liverpool. (P. Macmuldrow, Managing Director.)	Macmuldrow, Liverpool.
Civil Service Supply Association, Limited, 136, Queen Victoria Street, London	Stores, London.
Clabburn, James, Norwich	
Clay & Abraham, 87, Bold Street, Liverpool . . .	One, Liverpool.
Clay Cross Company, The, Clay Cross Collieries, near Chesterfield	Jackson, Clay Cross.
Clode, A. O., Letchmore Heath, Elstree	Clode, Radlett.
Coates & Co., Blackfriars Distillery, Plymouth. . .	Coates, Plymouth.
Cochrane, Paterson & Co., Leith	Cochrane, Leith.
Cochrane, Thos. K., Lieutenant, H. M. S. *Invincible*, Southampton.	
Collingridge, W. H. and L., City Press, 148 and 149, Aldersgate Street, London	Collingridge, London.
Collings, J.A., Monce Square, and Richmond Walk, Devonport	Jacoals, Devonport.
Colville, David, & Sons, 7, Fenchurch Avenue, London .	Colville, Motherwell.
Donacher, James, & Sons, Bath Buildings, Organ Works, Huddersfield	Ericht, Huddersfield.
Conan, Joseph, 4, Dawson Street, Dublin	Conan, Dublin.
Cook, Son & Co., 22, St. Paul's Churchyard, London . .	Cook, St. Paul's, London.
Coope, K. Jesser, S.S. Yacht *Sunrise*	
Corrie, William, & Co., 114, 116, 118, and 120, Cromac Street, Belfast	Cromac, Belfast.
Cotterell Brothers, 11, Clare Street, Bristol . . .	Cotterells, Bristol.
Couper, James, & Sons, City Glass Works, Glasgow . .	Coupers, Glasgow.
Coventry Cycle Company, White Friars Lane, Coventry .	Imperial, Coventry.
Coventry Machinists' Company, Limited, 15 and 16, Holborn Viaduct, London	Cheylsmore, London.
Cowie Brothers & Co., 59, St. Vincent Street, Glasgow .	Celtic, Glasgow.

NAME OF FIRM.	REGISTERED TELEGRAPHIC ADDRESS.
Crossley, John, & Sons, Limited:—	
Halifax: Dean Clough Mills	Crossleys, Halifax.
London: Falcon Hall, 15, Silver Street . . .	Crossleys Limited, London.
Manchester: 57, Portland Street	Tapestry, Manchester.
Cunard Steamship Co., Limited, Liverpool	Cunard, Liverpool.
Davey & Son, Old Barge House Wharf, Blackfriars Bridge, London	Wallsend, London.
Dawson, O. E., 10, Hart Street, Bloomsbury Square, London, W.C.	Dorart, London.
Debenham & Freebody, Wigmore Street and Welbeck Street, London	Debenham, London.
Debenham, Tewson, Farmer, & Bridgwater, 80, Cheapside, London	Debenhams, Cheapside, London.
Dickeson, Sir Richard, and Company, Victoria Warehouses, Mansell Street, London	Richard Dickeson, London.
Dover: Market Lane	Dickeson, Dover.
Dublin: Ellis's Quay	Dickeson, Dublin.
Aldershot	Dickeson, Aldershot.
Dickinson, John, & Co., 65, Old Bailey, London . . .	Commiles, London.
Dinning & Cooke, Percy Iron Works, Newcastle-on-Tyne .	Dinning, Newcastle-on-Tyne.
Downing, J. S., Crown Works, Commercial Street, Birmingham	Downing, Birmingham.
Dublin and Wicklow Manure Company, Limited, 1, College Street, Dublin	Dubwick, Dublin.
Dunn & Wright, Printers and Publishers, 100—102, West George Street and 100—106, Stirling Road, Glasgow .	Chardon, Glasgow.
Eason, Charles, Publisher, 13, Aston's Quay, Dublin .	Eason, Dublin.
Eglinton Chemical Co., Limited, Irvine, N.B. . . .	Eglinton, Irvine.
Emmet, John, & Co., Springfield Paper Mills, Bolton .	Emmet, Bolton.
English and Foreign Electrotype Agency, 19, Ludgate Hill, London, E.C.	Electragt, London.
Eyre & Spottiswoode, Great New Street, London . .	Spotless, London.
Faris, David, Warstone Lane, Birmingham . . .	Faris, Birmingham.
Faudel, Phillips & Sons, 36 to 40, Newgate Street, London	Faudel, London.
Fauvel, C. H., 4, Carlton Terrace, Southampton . .	
Feltham & Co., City Steam Works, Little Britain, London .	Felthams, London.
Fernau & Eltze, 133, West Campbell Street, Glasgow . .	Fernau, Glasgow.
Finlayson, Bousfield & Co., Flax Mills, Johnstone, N.B. .	Finlayson, Johnstone.
Fleming, F. & A. B. & Co., Limited, Chemical Works, Caroline Park, Edinburgh; (London Office, 15, Whitefriars St.)	Caroline, Edinburgh.
Foster, Porter & Co., Limited, 47, Wood Street, London .	Fosporter, London.
Fraser & Fraser, Bromley-by-Bow, London . . .	Pressure, London.
Fulcher, Arthur, Milgate Park, Maidstone. (Telegraph Station, Bearstead)	
Gardner, Jno., & Son, 11, Bradford Street, Birmingham .	Simplex, Birmingham.
Gillig's United States Exchange, 9, Strand, Charing Cross, London	Rendezvous, London.
Glasgow and West of Scotland Guardian Society for the Protection of Trade, 145, Queen Street, Glasgow . .	Guardian, Glasgow.
Dundee	Guardian, Dundee.
Godden, William Jefferys, Solicitor, Point Pinellas, Hillsborough County, Florida, U.S.A.	
Grand Hotel, Broad Street, Bristol . .	Grand, Bristol.
Grand Hotel, Trafalgar Square, London	Granotel, London.

H *

NAME OF FIRM.	REGISTERED TELEGRAPHIC ADDRESS.
Grant & Co., 72 and 75, Turnmill Street London . .	Grants, London.
Green & Sons, Printers, Beverley	Green, Beverley.
Green, P., & Co., 13, Fenchurch Avenue, London. . .	Thirteen, London.
Green, H. G. Egerton, King's Ford, Colchester . .	
Greener, W. W., St. Mary's Square, Birmingham. . .	Greener, Birmingham.
London : 68, Haymarket	Ejector, London.
Guion & Co., 11, Rumford Street, Liverpool . . .	Guion, Liverpool.
Haage & Schmidt, Erfurt	Hageamit, Erfurt.
Hallett & Co., 7, St. Martin's Place, London . . .	Olgrande, London.
Hall, J. & R., 23, St. Swithin's Lane, London . . .	Hallford, London.
Dartford, Kent	Hallford, Dartford.
Hansard, Henry & Son, House of Commons Printing Office, 41, Parker Street, W.C..	Hansards, London.
Haslam, John, & Co., Limited, Fountain Street, Manchester	Squirrels, Manchester.
Bolton	Squirrels, Bolton.
London	Squirrels, London.
Haynes, George, & Co., Hampstead Cotton Mills, Cherry Tree Lane, Stockport	Haynes, Stockport.
Henderson Brothers, "Anchor" Line, 47, Union Street, Glasgow	Anchor, Glasgow.
Henry, A. S., & Co., Huddersfield	Henrys, Huddersfield.
Heywood, John, Publisher and Bookseller, Deansgate, Manchester	Books, Manchester.
Hildesheimer & Faulkner, 41, Jewin Street, London . .	Labor, London.
Hobbs, W., & Sons, Printers and Chromo-Lithographers, Maidstone	Hobbs, Maidstone.
Hogg, Alexander, & Co., 60, Virginia Street, Glasgow .	Hogg, Glasgow.
Holden, Burnley & Co., Cumberland Works, Cemetery Road, Bradford	Burnley, Girlington.
Holden, George, & Son, 21, Carter Lane, Old Change, London	Holden, London.
Holden, Isaac, & Sons, Alston Works, Bradford . .	Holdens, Bradford.
Hooper & Co., Covent Garden, London	Hortus, London.
Horn & Son, Patent Agents, Somerset Chambers, 151, Strand, London	Wide-a-wake, London.
Hornbuckle, W. A., & Co., 18, Billiter Street, London. .	Hornbuckle, London.
Hôtel Métropole, Northumberland Avenue, London . .	Métropole, London.
Houlston & Sons, 7, Paternoster Buildings, London . .	Houlston, London.
Howatson, George S., 20, Bucklersbury, London . . .	George Howatson, London.
Howell, Henry, & Co., 180, Old Street, London . . .	Henry Howell, London.
Howell, John, & Co., Limited, 3, St. Paul's Churchyard, London	Howell, London.
Hughes, Thomas, & Co., 194, Euston Road, London . .	Navigable, London.
Illingworth, Daniel, & Sons, Whetley Mills, Bradford .	Illingworth, Bradford.
Ismay, John, & Sons, Newcastle-on-Tyne	Ismays, Newcastle-on-Tyne.
Jarvis, J. W., & Son, 28, King William Street, Strand, London	Biblionist, London.
Jennings, George, Palace Wharf, Stangate, London . .	Jennings, London.
Johnson, Matthey & Co., 78, Hatton Garden, London .	Matthey, London.
Johnson, Walker & Tolhurst, 80, Aldersgate Street, London	Jowato, London.
Johnston, W. and A. K., 5, White Hart Street, Warwick Lane, London	Geographers, London.

NAME OF FIRM.	REGISTERED TELEGRAPHIC ADDRESS.
Junior Army and Navy Stores, Limited :—	
London : York House, Waterloo Place . . .	Supplies, London.
Aldershot : 16, 17, and 18, Union Street . . .	Supplies, Aldershot.
Dublin : 22, 23, and 24, D'Olier Street . . .	Supplies, Dublin.
Keen, Robinson & Bellville, Garlick Hill, Cannon Street, London	Keen, London.
Kerr & Richardson, Wholesale Stationers, 89, Queen Street, Glasgow	Ellisland, Glasgow.
Keyser, A., & Co., Foreign Bankers, 21, Cornhill, London :—	
For Inland telegrams	Keyser, Cornhill, London.
For Foreign and Colonial telegrams	Keyser, London.
Kiddier, J., & Son, Globe Works, Waterway Street, Nottingham	Kiddier, Nottingham.
Kirby, Beard & Co., 115, Newgate Street, London . .	Ravenhurst, London.
Lang & Co., 146 and 150, Ingram Street, Glasgow . .	Langkelty, Glasgow.
London : 16, Watling Street	Langkelty, London.
Leadbeater & Scott, St. Mary's Works, Peniston Road, Sheffield	Leadbeater, Sheffield.
Leaf, Sons & Co., Old Change, London	Leaf, London.
Leckie, John, & Co., Walsall	Leckie, Walsall.
London : 12, St. Mary Axe	Saddlery, London.
Legbrannock District Collieries Company, Limited, 21, Hope Street, near Central Station, Glasgow . .	Waldie, Glasgow.
Leisler, Bock, & Co., Glasgow	Leisler, Glasgow.
Leith Distillery, Leith	Bernard, Leith.
Lepard & Smiths, 29, King Street, Covent Garden, London	Lepard, London.
Leslie, D., Wholesale Stationer, Perth	Leslie, Perth.
Letts, Charles, & Co., 3, Royal Exchange, London .	Diarists, London.
Lilleshall Company, Limited, Priors Lee Hall, near Shifnal	Lilleshall, Oakengates.
Lindley, C., & Co., 34, Englefield Road, London . . .	Beauvoir, London.
Linklater & Niven, Queen's Chambers, Pirie Street, Adelaide, South Australia	Haddon, Adelaide.
Lister & Co., Manningham Mills, Bradford . . .	Lister, Bradford.
Little, John, The Library, Wrexham	Little, Wrexham.
Lockwood (Crosby) & Co., 7, Stationers' Hall Court, Ludgate Hill, London	Crosblook, London.
London Mercantile Association, Limited, 8, Finch Lane, London	Mercable.
Lotz, Abbott & Co., 66, Queen Street, London . .	Ingbooth, London.
Maclure & Macdonald, 2, Bothwell Circus, Glasgow .	Lithographers, Glasgow.
Macmuldrow, Peter, Steam Saw Mills, London Road, Liverpool	Macmuldrow, Liverpool.
Macniven & Wallace, 132, Princes Street, Edinburgh .	Library, Edinburgh.
Maconochie Brothers, Lowestoft	Maconochie, Lowestoft.
London : 1, East India Avenue	Maconochie, London.
Malton Gas Company, Gas Works, Malton . . .	Gas, Malton.
Mardon, Son & Hall, The Caxton Works, Bristol .	Mardon's, Bristol.
Marlborough, R., & Co., 51, Old Bailey, London . .	Marlborough, London.
Marlborough, Gould & Co., 52, Old Bailey, London .	
Marquess of Ailsa	Ailsa, Maybole.
Marriott, H. & F. A., Birstall, near Leeds . . .	Marriott, Birstall.
Marshall Brothers, 3, Amen Corner, Paternoster Row, London	Grapho, London.

NAME OF FIRM.	REGISTERED TELEGRAPHIC ADDRESS.
Mawson, Swan & Morgan, Stationers, &c., Newcastle-on-Tyne.	Morgan, Newcastle-on-Tyne.
Merry & Cuninghame, 127, St. Vincent Street, Glasgow	Merry, Glasgow.
Meux & Co., Limited, Horse Shoe Brewery, London	Meuxs, London.
Midland Educational Company, Limited, 91 and 92, New Street, Birmingham	Educational, Birmingham.
Midland Grand Hotel, St. Pancras Station, London	Midotel, London.
Millington & Sons, 32, Budge Row, London	Millington, London.
Montagu, Samuel, & Co., 60, Old Broad Street, London	Montagu, London.
Morris & Griffin, Ceres Works, Wolverhampton	Ceres, Wolverhampton.
Newport, Monmouth	Fuller, Maindee, Mon.
Morrison & Gibb, 11, Queen Street, Edinburgh	Magazine, Edinburgh.
Mort, Liddell & Co., Widnes	Mort, Widnes.
Mortimer, Edward, Bookseller and Printer, Halifax	Mortimer, Halifax.
Motherwell, R., & Co., 43, Queen Street, Glasgow	Motherwell, Glasgow.
Mowbray, A. R., & Co., 1186, St. Aldate's, Oxford	Mowbray, Oxford.
London : 65, Farringdon Street	Anglicanus, London.
Nettlefold & Sons, 54, High Holborn, London	Nettleson, London.
Newton, Chambers & Co., Limited, Thorncliffe Iron Works and Collieries, near Sheffield	Newton, Sheffield.
Normandy, A. Stilwell, & Co., Victoria Docks, London	Normandy, London.
Northcote, Stafford, & Co., St. Paul's Churchyard, London	Stafford Northcote, London.
North of England School Furnishing Co., Limited, 25, Grainger Street West, Newcastle-on-Tyne	Scholastic, Newcastle-on-Tyne.
Novelli & Co., Billiter House, Billiter Street, London	Novelli, London.
Ocean and Continental Express, 35, Haymarket, London	Nixon, London.
Oetzmann & Co., 67 to 79, Hampstead Road, London	Oetzmann, London.
Ogilvie & Moore, Warren's Place, Cork	Ogilvie, Cork.
Orient Steam Navigation Co., 'Limited, 13, Fenchurch Avenue, London	Orient, London.
Page & Sandeman, 5½, Pall Mall, London	Torniport, London.
Pantin, W. & C., 147, Upper Thames Street, London	Pantinko, London.
Partridge, S. W., & Co., 9, Paternoster Row, London	Pictorial, London.
Paul (Kegan), Trench & Co., 1, Paternoster Square, London	Columnae, London.
Pawson & Co., Limited, St. Paul's Churchyard, London	Pawson, London.
Peake, Thomas, The Tileries, Tunstall, Staffordshire	Peake, Tunstall, Staff.
Philip, Son & Nephew, 51, South Castle Street, Liverpool	Education, Liverpool.
Pigou, Wilks & Laurence, Limited, 11, Queen Victoria Street, London	Pigou, London.
Pirie, Alex., & Sons, Limited, Stoneywood Works, Auchmill, Aberdeenshire	Pirie, Aberdeen.
Pitman, Frederick, 20 and 21, Paternoster Row, London	Pitman, London.
Pitt & Scott, Foreign Parcels Express :—	
London : 23, Cannon Street	Pitt Scott, London.
Liverpool : Corf's Buildings, 16, Preeson's Row	Pitt Scott, Liverpool.
Paris : 7, Rue Scribe	Pitt, Paris.
Poole, Henry, & Co., 36 to 39, Savile Row, London	Eloop, London.
Porteous, James & Son, 5, Dixon Street, Glasgow	Porteous, Glasgow.
Poulten, Thomas, & Sons, 6, Arthur Street West, London Bridge, London	Poulter, London.
Price, R. J. Lloyd, 10, Wilton Crescent, London	Canis, London.
Rhiwlas, Bala, North Wales	Price, Bala.

NAME OF FIRM.	REGISTERED TELEGRAPHIC ADDRESS.
Price, Sons, & Co., Bristol	Price, Bristol.
Pullman Company, Limited, St. Pancras Station, London .	Pullman, London.
Pullman, R. & J., 17, Greek Street, Soho, London . .	Leathersellers, London.
Religious Tract and Book Society of Scotland, Edinburgh	Tract Society, Edinburgh.
Remington & Co., 18, Henrietta Street, Covent Garden, London . . . . }	Proficio, London.
Richardson & Chadbaum, 8, Finch Lane, London . .	Mercable, London.
Royal Mail Steam Packet Co., 18, Moorgate Street, London	Omarius, London.
Rylands & Sons, Limited, New High Street, Manchester .	Rylands, Manchester.
Samuel & Escombe, 26, Austin Friars, E.C. . . .	Gainsay, London.
Selwood Printing Works, Frome	Selwood, Frome.
Shelton Iron & Steel Company, Limited, Stoke-upon-Trent, Staffordshire . . . . }	Shelton, Stoke-on-Trent.
London Offices : 122, Cannon Street, E.C. . .	Sheltonian, London.
Sherwill, J. H., Market Street, Devonport	Sherwill,Grocer,Devonport.
Siemens Brothers & Co., Limited, 12, Queen Anne's Gate, London . . . . }	Siemens, London.
Silverlock, Henry, 92, Blackfriars Road, London . .	Silverlock, London.
Smith & Ebbs, Northumberland Buildings, Fenchurch Street, London . . . . }	Adept, London.
Smith, W. H., & Son, 80, Middle Abbey Street, Dublin .	Season, Dublin.
Sotheran, Henry, & Co., 136, Strand, London . . . }	Bookmen, London.
London : 36, Piccadilly	
Manchester : 49, Cross Street	Bookmen, Manchester.
Squire & Sons, 413, Oxford Street, London	Squire, London.
Stanhope Company, Limited, 20, Bucklersbury, London .	Stanhope, London.
Stapley & Smith, London Wall, London	Stapley, London.
Star Brush Company, The, Eden Grove, Holloway, N. .	Stellatus, London.
Starkey, R. W., & Son, 27, New Bridge Street, Blackfriars, London . . . . }	Starkey, London.
Steel Company of Scotland, Limited, 150, Hope Street, Glasgow . . . . }	Steel, Glasgow.
Stone, Henry, & Son, Box Manufactory, Banbury . .	Stone, Banbury.
Stubbs' Mercantile Offices, 42, Gresham Street, London .	Stubbs, London.
Summerlee & Mossend Iron & Steel Company, Limited, 172, West George Street, Glasgow . . . . }	Summerlee, Glasgow.
Swiss, A. H., Bookseller and Printer, Devonport . .	Alfred Swiss, Devonport.
Tate, Henry, & Sons :—	
London : 21, Mincing Lane	Tateson, London.
Silvertown : Thames Sugar Refinery . . - .	Tate, Silvertown.
Liverpool : 15H, Exchange Buildings	Tateson, Liverpool.
Terrell, William, & Sons, Limited, Welsh Back, Bristol .	Terrell, Bristol.
Tharsis Sulphur and Copper Co., Limited, 136, West George Street, Glasgow . . . . . . }	Tharsis, Glasgow.
Thomson, Henry, & Co., Newry	Thomson, Newry.
Thomson & Campbell, 5, Adelphi Terrace, Strand . .	Yacht, London.
Tuck, Raphael, & Sons, 72 and 73, Coleman Street, London	Palette, London.
Union Steamship Company, Limited, 11, Leadenhall Street, London . . . . }	Oregon, London.
United Asbestos Co., Limited, 161, Queen Victoria Street, London . . . . }	Asbestos, London.
Unwin, Robert, & Co., 1, Old Hall Street, Liverpool . .	Silverstone, Liverpool.
Vicars, T. and T., Seel Street, Liverpool	Cutters, Liverpool.
London : 20, Bucklersbury	George Howatson,London.

NAME OF FIRM.	REGISTERED TELEGRAPHIC ADDRESS.
Vulcan Iron Works, Langley Mill, near Nottingham .	Turner, Langley Mill.
Walch & Butler, Solicitors, Hobart, Tasmania . . .	Vibex, Hobart.
Walker, Howard & Co., 70, Lower Thames Street, London .	Evering, London.
Walker, John, & Co., Farringdon House, Warwick Lane, London . . . . }	Chebucto, London.
Waterlow & Sons, Limited, London Wall, London . .	Waterlow Sons, London.
Waterston, George, & Sons, 56, Hanover Street, Edinburgh	Waterstons, Edinburgh.
London : 9, Rose Street, Newgate Street . . .	Waterstons, London.
Watson & Co., G. L., 108, West Regent Street, Glasgow .	Vril, Glasgow.
Wells, A., & Co., Steam Works, Spanish Road, Wandsworth, London . . . . }	Tinkery, London.
Werner & Pfleiderer, 86, Upper Ground Street, Blackfriars Bridge, London . . . . }	Pfleiderer, London.
White, John, 26, Great St. Helens, London	John White, London.
White, Robert Stanley, Solicitor, 12, New Inn, Strand, London, W.C.; and Queen Anne's Lodge, Lordship Road, Stoke Newington.	
Widnes Alkali Co., Limited, Widnes	Widnes, Widnes.
Liverpool : 1 and 2, Bank Buildings, 60, Castle Street	Soda, Liverpool.
Williams, J. D., & Co., Langley Mills Manufacturing Co., Manchester . . . . }	Witches, Manchester.
Wilmott, Edward W., Passenger Shipping Agent, Malta .	Wilmott, Malta.
Winn & Holland, Montreal	Winn, Montreal.
Wood & Ingram, The Nurseries, Huntingdon . . .	Ingram, Huntingdon.
Wood's Hotel, Furnival's Inn, London	Woodsdon, London.
Woodward, Clark, & Co., Nottingham	Woodward, Nottingham.
Worthington & Co., Brewers, Burton	Worthingtons, Burton.
Wright, Layman, & Umney, 50, Southwark Street, London	Umney, London.
Wrigley, James, & Son, Limited :—	
London : 21, Budge Row	Wrigleys, London.
Bury	Wrigleys, Bury.
Manchester	Wandson, Manchester.

PRINTED BY CASSELL & COMPANY, LIMITED, LA BELLE SAUVAGE, LONDON, E.C.

10.1080

www.ingramcontent.com/pod-product-compliance
Lightning Source LLC
Chambersburg PA
CBHW021822190326
41518CB00007B/704